Environmental Futures

Journal of the Royal Anthropological Institute Special Issue Series

The Journal of the Royal Anthropological Institute is the principal journal of the oldest anthropological organization in the world. It has attracted and inspired some of the world's greatest thinkers. International in scope, it presents accessible papers aimed at a broad anthropological readership. We are delighted to announce that from 2014 the annual special issues will also be available from the Wiley Blackwell books catalogue.

Previous special issues of the JRAI:

Doubt, Conflict, Mediation: The Anthropology of Modern Time, edited by Laura Bear

Blood Will Out: Essays on Liquid Transfers and Flows, edited by Janet Carsten

The Return to Hospitality: Strangers, Guests, and Ambiguous Encounters, edited by Matei Candea and Giovanni da Col

The Aesthetics of Nations: Anthropological and Historical Approaches, edited by Nayanika Mookherjee and Christopher Pinney

Making Knowledge: Explorations of the Indissoluble Relation between Mind, Body and Environment, edited by Trevor H.J. Marchand

Islam, Politics, Anthropology, edited by Filippo Osella and Benjamin Soares

The Objects of Evidence: Anthropological Approaches to the Production of Knowledge, edited by Matthew Engelke

Wind, Life, Health: Anthropological and Historical Perspectives, edited by Elisabeth Hsu and Chris Low

Ethnobiology and the Science of Humankind, edited by Roy Ellen

The Power of Example: Anthropological Explorations in Persuasion, Evocation, and Imitation, edited by Andreas Bandak and Lars Højer

ENVIRONMENTAL FUTURES

EDITED BY JESSICA BARNES

This edition first published 2016
© 2016 Royal Anthropological Institute

Registered Office
JohnWiley & Sons Ltd, The Atrium, Southern Gate, Chichester, West Sussex, PO19 8SQ, UK

Editorial Offices
350 Main Street, Malden, MA 02148-5020, USA
9600 Garsington Road, Oxford, OX4 2DQ, UK
The Atrium, Southern Gate, Chichester, West Sussex, PO19 8SQ, UK

For details of our global editorial offices, for customer services, and for information about how to apply for permission to reuse the copyright material in this book, please see our website at www.wiley.com/wiley-blackwell.

The right of Jessica Barnes to be identified as the author of the editorial material in this work has been asserted in accordance with the UK Copyright, Designs and Patents Act 1988.

Library of Congress Cataloging-in-Publication Data

Environmental Futures: Edited by Jessica Barnes
 pages cm. – (Journal of the Royal Anthropological Institute special issue series)
 Includes bibliographical references and index.
 ISBN 9781119278320 (alk. paper)
 CIP requested

 9781119278320

A catalogue record for this book is available from the British Library.

Journal of the Royal Anthropological Institute.
Incorporating MAN
Print ISSN 1359-0987
All articles published within this special issue are included within the ISI Journal Citation Reports® Social Science Citation Index. Please cite the articles as volume 22(Supp) of the Journal of the Royal Anthropological Institute.

Cover image: *Romulus consulting the augury*, by John Leech. In Beckett, G., *The comic history of Rome* (Evans & Co, 1850s), http://posner.library.cmu.edu/Posner/books/book.cgi?call=937_A138C_1850. Photo by Andreas Praefcke.

Cover design by Ben Higgins

Set in 10 on 12pt Minion by Aptara Inc.

Printed in Singapore by C.O.S. Printers Pte Ltd

1 2016

Contents

Notes on contributors   vi

1   ANDREW S. MATHEWS & JESSICA BARNES *Prognosis: visions
    of environmental futures*   9

2   JESSICA O'REILLY *Sensing the ice: field science, models, and
    expert intimacy with knowledge*   27

3   JESSICA BARNES *Uncertainty in the signal: modelling Egypt's
    water futures*   46

4   DAVID KNEAS *Subsoil abundance and surface absence: a
    junior mining company and its performance of prognosis in
    Northwestern Ecuador*   67

5   NUSRAT SABINA CHOWDHURY *Mines and signs: resource and
    political futures in Bangladesh*   87

6   KAREN HÉBERT *Chronicle of a disaster foretold: scientific risk
    assessment, public participation, and the politics of
    imperilment in Bristol Bay, Alaska*   108

7   GISA WESZKALNYS *A doubtful hope: resource affect in a
    future oil economy*   127

8   MANDANA LIMBERT *Liquid Oman: oil, water, and causality
    in Southern Arabia*   147

9   AUSTIN ZEIDERMAN *Prognosis past: the temporal politics of
    disaster in Colombia*   163

10 ELIZABETH FERRY *Claiming futures*   181

Index   189

Notes on contributors

Jessica Barnes is an Assistant Professor in the Department of Geography and the Environment & Sustainability Program at the University of South Carolina. She is author of *Cultivating the Nile: the everyday politics of water in Egypt* (Duke University Press, 2014) and editor (with Michael R. Dove) of *Climate cultures: anthropological perspectives on climate change* (Yale University Press, 2015). *Department of Geography, University of South Carolina, Callcott Building, 709 Bull St, Columbia, SC 29208, USA. jebarnes@mailbox.sc.edu*

Nusrat Sabina Chowdhury is an Assistant Professor of Anthropology at Amherst College. She is now completing a book on crowds and protest in Bangladesh based on fieldwork in Phulbari in the northwest and in Dhaka, the capital. *Department of Anthropology and Sociology, Amherst College, 203B Morgan Hall, PO Box: AC# 2226, Amherst, MA 01002, USA. nchowdhury@amherst.edu*

Elizabeth Ferry is a Professor of Anthropology at Brandeis University, Waltham, Massachusetts. She is author of *Not ours alone: patrimony, collectivity and value in contemporary Mexico* (Columbia University Press, 2005) and *Minerals, collecting, and value across the US-Mexican border* (Indiana University Press, 2013), and editor (with Mandana Limbert) of *Timely assets: the politics of resources and their temporalities* (School of Advanced Research Press, 2008). She is currently working on a book on perceptions of physical gold and other gold-based assets in mines and financial markets. *Department of Anthropology, Brandeis University, Brown 228, 415 South Street, Waltham, MA 02453, USA. ferry@brandeis.edu*

Karen Hébert is an Assistant Professor jointly appointed in the Department of Anthropology and the School of Forestry & Environmental Studies at Yale University. She is currently completing a book that examines intersecting issues of environment and economy in coastal Alaska, which will be published by Yale University Press. *Department of Anthropology, Yale University, 10 Sachem Street, New Haven, CT 06511, USA. karen.hebert@yale.edu*

David Kneas is an Assistant Professor in the Department of Geography and the Environment & Sustainability Program at the University of South Carolina. He is currently completing a book that explores the histories of and tensions between material and imagined resources in the Ecuadorian Andes. *Department of Geography, University of South Carolina, Callcott Building, 709 Bull Street, Columbia, SC 29208, USA. kneas@mailbox.sc.edu*

Mandana Limbert is an Associate Professor in the Department of Anthropology in the Queens College and the Graduate Center of the City University of New York. She is author of *In the time of oil: piety, memory, and social life in an Omani town* (Stanford University Press, 2010) and editor (with Elizabeth Ferry) of *Timely assets: the politics of resources and their temporalities* (School for Advanced Research Press, 2008). *Department of Anthropology, Queens College, Powdermaker Hall 314, 65-30 Kissena Blvd, Flushing, NY 11367, USA. Mandana.Limbert@qc.cuny.edu*

Andrew S. Mathews is an Associate Professor of Anthropology at the University of California, Santa Cruz. He is author of *Instituting nature: authority, expertise and power in Mexican forests* (MIT Press, 2011). *Department of Anthropology, University of California, Santa Cruz, 1156 High Street, Santa Cruz, CA 95064, USA. amathews@ucsc.edu*

Jessica O'Reilly is an Assistant Professor of Anthropology at the College of Saint Benedict and Saint John's University. She is author of *The technocratic Antarctic: an ethnography of scientific expertise and environmental governance* (Cornell University Press, forthcoming). *Sociology Department, College of Saint Benedict and Saint John's University, 2850 Abbey Plaza, Collegeville, MN 56321, USA. jloreilly@csbsju.edu*

Gisa Weszkalnys is an Assistant Professor of Anthropology at the London School of Economics and Political Science. She is author of *Berlin, Alexanderplatz: transforming place in a unified Germany* (Berghahn, 2010) and editor (with Simone Abram) of *Elusive promises: planning in the contemporary world* (Berghahn, 2013). *Department of Anthropology, London School of Economics and Political Science, Houghton Street, London WC2A 2AE, UK. g.weszkalnys@lse.ac.uk*

Austin Zeiderman is an Assistant Professor of Urban Geography at the London School of Economics and Political Science. He is author of *Endangered city: the politics of security and risk in Bogotá* (Duke University Press, 2016). *Department of Geography & Environment, London School of Economics and Political Science, Houghton Street, London WC2A 2AE, UK. A.Zeiderman@lse.ac.uk*

1

Prognosis: visions
of environmental futures

ANDREW S. MATHEWS *University of California, Santa Cruz*

JESSICA BARNES *University of South Carolina*

While prognoses about the future are as old as human society, this special issue argues that the proliferation of new ways of modelling, planning, and interpolating the future of resources and environments is an increasing feature of contemporary environmental politics. In our introduction, we draw out two dimensions to this prognostic politics: first, the processes of making predictions about the future; and second, the movement of these predictions through the unstable and messy institutions that act upon the future in the present. We argue that new regimes of environmental forecasting and contests over these prognoses are giving rise to new forms of nature, framings of time and space, and modes of politics.

In the contemporary world, various environmental futures, from the apocalyptic to the utopian, are brought to us by a host of people and institutions. Stories about global climate change, ocean acidification, biodiversity loss, oil depletion, water scarcity, and deforestation conjure a picture of imminent doom, while counter-visions of organic agriculture, solar panels, rehabilitated ecosystems, and green cities vie to suggest more hopeful, alternative futures. Often, proponents of economic or technological change step forward to speak for futures that are, on the one hand, inevitable and desirable, and, on the other hand, in need of care if they are to come into being. Across domains of artistic, technological, and popular culture, efforts to imagine and domesticate these futures proliferate.

Concern about the future is nothing new. Humans have always struggled to imagine and predict what is to come, and future-orientated practices are as old as human society. We could point, for example, to the recourse to augury or oracles in ancient civilizations (Herodotus 1987) or the long traditions of divination in many parts of the world (Evans-Pritchard 1963 [1937]; Holbraad 2012). Modern ways of addressing the future developed alongside the emergence of modern states and conceptions of a demarcation between science, politics, and religion (Shapin & Schaffer 1985), and constructions of the secular and the non-secular (Lilla 2008). Central to modern statecraft, therefore, was the generation of a set of practices for dealing with the uncertainty that surrounds the future, for instance through statistical modes of reasoning (Desrosières 1991; Hacking 1990) and forms of risk management and social insurance (Beck 1992). Furthermore, just as states have sought to manipulate the past by celebrating some moments in history and silencing others, so, too, they have linked these selected pasts to particular futures. The question of the past, then, if looked at in the right way, has always also been a question about the future.

It is important to remember that religion, too, has been an important source of imagination and action around environmental futures, often by those who redefine or deny altogether the boundaries between the scientific, the political, and the religious. Examples include the visions of environmental futures held by conservative Christian Evangelicals opposed to state-sanctioned climate science in the United States (McCammack 2007) or the futures imagined by indigenous people concerned about the protection of sacred mountains (de la Cadena 2010). While the papers in this volume do not address these non-secular visions of the future, there are numerous examples in the world, and there are likely to be many more as anthropologists probe further the environmental futures imagined by non-scientists.

Within anthropology, there has been a recent upsurge of interest in thinking about the future.[1] With some notable exceptions (e.g. Wallman 1992), much of the earlier anthropological work on temporalities focused on the past and present, and the past in the present, with a relative neglect of the future (Munn 1992: 115-16). A growing number of anthropologists have, however, begun to explore the future as a domain which, like the past, can be brought into the present to do various kinds of political or cultural work (Abram & Weszkalnys 2013; Appadurai 2013; Bear 2014; Ferry & Limbert 2008; Guyer 2007; Holbraad & Pedersen 2013; Maurer 2002; Miyazaki 2006; Rosenberg & Harding 2005). This scholarship has drawn attention to the state of anticipation, as something that 'pervades the way we think about, feel and address our contemporary problems' (Adams, Murphy & Clarke 2009: 248).

Although interest in the future is not novel, we believe that we are seeing the emergence and proliferation of new ways of thinking about the future, and new ways of linking the future with the present or the past. As the essays in this special issue show, modelling, planning, and interpolating the future of resources and environments has become an increasing feature of contemporary environmental politics. Rather than operating alone, these new methods of describing the future are in conversation with pre-existing technical and political arts of imagining and acting upon the future, such as national estimates of water, fisheries, oil, or forests. In the deployment of these environmental futures, we see a reworking of the relationship between states, corporations, and their publics, which opens up new points for opposition or engagement on the part of the people who might be affected by governmental or corporate actions.

In the remainder of this introduction, we draw out two dimensions to this prognostic politics. First, we look at the processes of making predictions about the future. Second, we look at the movement of these predictions through the unstable and messy institutions and publics that act upon the future in the present. These two dimensions are not distinct. As various actors conduct their predictive work, they are haunted by the political implications of their results, and their imaginations are shaped by interactions with different publics. So too, as predictions move, these circulations rework the content of models and the practices of modellers, with multiple interactions at various stages, and no simple flow from a new future to a new policy. It is not the case, then, that powerful actors disseminate visions of the future to a passive and unknowing society (Wynne 1996). On the contrary, futures are everywhere remade, whether in laboratories by scientists who fear the political influence of an ignorant population (Wynne 2005), or in government offices by officials who are wary of public opinion, or on the streets by activists charting their own visions of what the future could be. Thus in the process of knowing, interpreting, and acting on the future, we see a close interaction between 'expert' and 'popular' domains. Indeed, citizen imaginations of the future may totally transform or constrain scientists' visions and the degree to which their ways of knowing the future gain traction and validity. Popular stories about the future, gossip about the credibility of official predictions, and accounts of conspiracies between officials and multinationals can all sustain or undermine official and scientific visions of environmental futures.

Knowing futures

The future in question may be a presence (e.g. a prospective mineral deposit), an absence (e.g. the end of oil), or a change (e.g. in prices or ice extent). This future may be anticipated, forecast, predicted, projected, prognosticated, divined, speculated, imagined, narrated, promised, revealed, augured, foreseen, or fantasized about. In casual conversation, we use these verbs of knowing the future interchangeably, sometimes switching from one to another without being fully aware of this change ourselves, despite the different implications that these terms carry (Weszkalnys 2014). Prognoses about the future can take place at various spatial and temporal scales, from the imminent to the long term (Guyer 2007). The future can be told as a story, calculated as a probability, or speculated upon as a form of potentially valuable risk. Those who make and interpret modelled futures struggle with ways to qualify it by using adjectives or visual imagery (Liverman 2009). Terms such as 'likely', 'probable', or 'expected' are used, sometimes in very specific technical ways, sometimes in more vernacular ways, and often with a curious slippage between the two by scientists themselves (Lahsen 2005). The future, then, is not one but many, and those who create futures typically seek to narrow down what the future can be to a relatively limited subset of possible registers.

   The texture of the future, like other facts, depends on how and by whom it is composed (Latour 2004). Sometimes, the biographies of particular futures affect their reception for an extended period of time. The credibility of futures of abundant and cheap nuclear energy, for example, was powerfully affected by the emergence of atomic energy in association with military research and the Cold War state (Jasanoff & Kim 2009), even as the Cold War also provided the material and ideological basis for studying global environmental change (Masco 2010). It is important, therefore, to look at who produces particular predictions about the future, and the historical contexts and institutional ecologies in which these futures emerged. The actors engaged in creating environmental futures range from computer modellers forecasting global climate patterns (Lahsen 2005), to farmers forecasting seasonal rainfall (Orlove, Chiang & Cane 2002), local residents forecasting the damages associated with an energy project (Howe 2014), and consultants forecasting natural gas prices (Mason 2007). In crafting their predictions, these actors draw both on their political imaginations about institutions and what they might do, and on an array of material engagements.

Journal of the Royal Anthropological Institute (N.S.), 9-26
© Royal Anthropological Institute 2016

Such material engagements feature in a number of the papers of this volume, from water dripping off an ice sheet (O'Reilly) to hard disks of data transported around the world (Barnes), rock-core samples stored in wooden boxes (Kneas), or slogans written on a house wall (Chowdhury).

By looking at the process of knowing the future, we hence focus on what kinds of knowledge and non-knowledge about the future are being made, who makes this knowledge (consultants, governments, scientists, computer programmers, 'locals'), how they make it (different calculative technologies, forms of data collection, and standardization practices), and what spatial and temporal scales they seek to grasp (the time periods they model, the landscapes they gather data from). Although in practice there is an enormously diverse range of methods for crafting the future, for the purposes of this volume we distinguish here three main aspects of the predictive process: model-building, the role of scenarios, and methods for dealing with uncertainty or risk.

Models

The emergence of techniques for modelling very complex systems, whether the global climate, ecosystems, economies, or markets, has been fundamental to allowing the proliferation of futures within contemporary environmental politics. These models draw upon vast quantities of empirical data, linking practices of measurement, standardization, and data entry, and connecting distant times and places (Edwards 2006). A 'culture of prediction' has evolved across many research fields, associated with the increasing availability of cheap desktop computing power and scientists' newfound ability to employ simulation models on an everyday basis (Johnson & Lenhard 2011). As the potential grows for models to describe what the environment is and how it is likely to change in the future, so too do demands for governments and other institutions to make themselves responsible for the natural resources, environments, or risks that have been revealed at new scales (Jasanoff 2004a; 2004b). Such demands translate, also, into critiques of these institutions for their failure to have met this responsibility in the past (Zeiderman, this volume).

Models have different purposes, face different material and infrastructural constraints, and are addressed to different audiences, with very different effects. The most charismatic and obvious of these models today are the general circulation models (GCMs) which simulate global climate patterns and enable the projection of climate change (Edwards 2010). These computer models, and the scientists who write the code and equations that comprise them, might be what first come to mind when we think of charting environmental futures. There are, however, many ways of modelling the future. Kirsten Hastrup (2013), for instance, describes the 'diagrammatic reasoning' that hunters in Greenland employ to comprehend the melting of ice around them. Operating through networks and images rather than concepts and numbers, this modelling is the hunters' way of reading the ice and anticipating both the near and more distant future. The kinds of models that have been built by natural resource managers are different again, relying upon practices of classifying, counting, and calculating the presence and possible growth of the resource in question, in order to plan the amount of the resource to be exploited. Models of tree growth, for example, emerged with the invention of scientific forestry at the end of the eighteenth century (Scott 1998). Such sustained-yield models are as much about performing the rationality and stability of the nation-state as they are about actually predicting and managing forests (Mathews 2011). Modelling, then, is found in many domains, and models can speak in several registers

Journal of the Royal Anthropological Institute (N.S.), 9-26
© Royal Anthropological Institute 2016

at once, perhaps asserting political authority and the stability of the nation-state, or perhaps predicting the quantified or non-quantified aspects of the future destiny of that resource.

The practices of modellers are deeply influenced by their political circumstances and biographies, the material resistances that they encounter in their daily work, and the audiences for whom they are building the model. It has long been observed by the scholarship on climate modelling that models are under-constrained, requiring experienced practical judgement and imagination on the part of modellers who convert equations to code and parameterize the models, inserting quantities or simple equations for domains that are too complex to model (Guillemot 2010; Sundberg 2009). Further, a model has to be 'tuned' through a skilled craft of adjusting its components so that it fits modellers' understandings and desires of what a 'good model' looks like. As Paul Edwards notes, 'better' may mean that the result agrees more closely with observations, or that it corresponds more closely to the modeller's judgement of what kind of change is physically plausible (2010: 342). This work of tuning is affected by modellers' disciplinary training and by their engagements with models and the subjects of those models, which are both intimate and technical, as we see in Jessica O'Reilly's paper on the glaciologists who study the future of the West Antarctic Ice Sheet. Modellers' decisions are affected, also, by their sense of the political environment (Lahsen 2009; van der Sluijs, Shackley & Wynne 1998).

Not all models are equally good for all users, and new models must typically be articulated with older models through a process of translation and commensuration which demonstrates the validity of the new model. What travels from one model to another may be quantitative data, or it may be only a sense of the plausible or reasonable limits and constraints that should be imposed upon the new model. Often, models must fit certain restrictions produced by other institutions or actors, such as the rules about how risk models are constructed owing to their effect on stock-market valuations (MacKenzie 2007). In the case of resource estimates, David Kneas's paper shows the manoeuvrings that a mining company has to do to bring its estimates of copper abundance in the Ecuadorian Andes in line with the rules set by the Toronto Stock Exchange.

Scenarios

Scenarios are a specific form of addressing the future which grew out of Cold War military planning, and have since been adopted across many social worlds. Scenarios require modellers to envisage a credible future which is, however, intrinsically impossible to calculate. During the Cold War, such imaginations would have been about the likely responses of enemy military leaders to a particular event or course of action. In the contemporary period, examples include the crafting of disciplined responses to bioterror events (Cooper 2006; Masco 2013), forest loss (Mathews 2015), financial market turns (Cooper 2010), or disease outbreaks (Lakoff 2008; Samimian-Darash 2013). Scenarios introduce a crucial new relation to an uncertain future event. Under scenario planning, a disciplined practice of imagining *plausible* future events enables concrete planning in the present to prepare for such events.

To formulate a scenario, it is necessary to select a set of reference-points, modes of observation, and objects of discourse (Mason 2007). This process is anchored in assessments by scenario planners as to what futures are credible. Like other stories, scenarios do the work of classifying which agents and objects are to be considered, where the story is to begin and end, and what kinds of major narrative structures are to be employed (Cronon 1992). Scenario planning is therefore a form of subjunctive narration about the future, a speculation of the form of 'What would happen if *x* happened?' We see this, for example, in the policies to reduce greenhouse gas emissions from forest removal, known as REDD (Reduced Emissions from Deforestation and Degradation). In these programmes, forest policies and reimbursements to landowners are justified by a 'reference' or 'business as usual scenario', which would have happened had the policy intervention not taken place (Mathews 2015). These scenarios of landcover change do not replace national statistical estimates of forest cover or growth; rather, they build upon and complement each other. Scenario planning has multiplied into different subfields, sometimes with the goal of optimizing policies (as with many climate change models), at other times with the goal of supporting vision-building or strategic choices (Westhoek, van den Berg & Bakkes 2006).

Scenario planning practices are potentially unstable and subject to contestation, or to slippage between different forms of scenario-making, from vision-building to policy optimization. In Karen Hébert's paper, for instance, she describes the controversy over the mining scenarios that the US Environmental Protection Agency used as the basis of its risk assessment for the Pebble Mine in Alaska. Hébert traces how the language of 'hypothetical mine scenarios' in the initial drafts of the assessment report was replaced with a language of 'realistic mine scenarios' in the final report. Such linguistic changes, she argues, have implications for how the possibility of a deviation from those scenarios is understood.

Risk, uncertainty, and disaster

When looking towards environmental futures, however, there will always be things that we do not and cannot know (although, as pointed out by Nelson, Geltzer & Hilgartner [2008], there are many dimensions of present environments about which we similarly know little). A key distinction in preparing to act upon the future is the difference between addressing *uncertainty*, the probability of an event whose nature is largely known, and *indeterminacy*, an event whose nature is not yet known. Preparations for disaster and risks of various kinds wrestle with both uncertain and indeterminate futures.

The evaluation of the risks that are associated with different environmental futures has come to be a key element in the prognosis process.[2] Risk analysis underlies the majority of modern environmental regulations, with more risky types of hazards being more tightly regulated (Jasanoff 1999). Typically, risk analysis has tried to delimit the type of future event that is calculated and prepared for, often by excluding non-expert visions of risk and causation, as, for example, when experts explain that they are responsible for the facts, and that ordinary citizens may only express ethical values and emotions. Indeed, some methodologies of risk assessment, such as the expert elicitation panels that Jessica O'Reilly discusses in her paper, are predicated on the notion that certain experts are in a position to evaluate future risks in a way that the 'average' person is not. More recently, however, as Karen Hébert's paper documents, there have been efforts to move towards more participatory processes in risk assessment, with mixed results.

When disasters do occur, there is often a reshuffling of systems of measurement, and assessments of causation, so that it becomes clear in retrospect what the nature of the risk was all along (Bond 2013). In other words, state institutions try to make risk tractable and calculable, both prospectively and retrospectively, as a way of demonstrating their reasonableness and competence. Such demonstrations are powerfully affected by popular understandings of risk and expertise, which can undermine or overthrow institutions at moments of spectacular failure (Jasanoff 2005; Zeiderman, this volume). As with other forms of modelling, then, risk models which seek to calculate quantifiable futures are sustained or undermined by other ways of assessing the future and the past, including collective understandings of the proper way of performing expert knowledge, or of the proper role of the state with regard to its audiences.

In response to the emergence of non-quantifiable forms of risk, the concept of reputational risk has come to the fore in the worlds of business, where the fact of having carried out a risk analysis is used to demonstrate responsibility and due diligence (Power 2007). In many cases, former regimes of statistical prediction have been displaced by risk-management control systems. Risk analysis is used in some measure to inocculate institutions against the environmental futures that they are supposed to anticipate in their management practices, by transforming a disaster from a demonstration of incompetence and lack of preparation, into something which was at least partially prepared for.

Futures in motion

Once environmental futures come into being, they are translated and transformed as they move through different sites. In the process, they produce a range of affective responses. In Gisa Weszkalnys's paper, for example, we see the 'doubtful hope' sparked by the future of oil in São Tomé and Príncipe, as well as the problematization of this affect by various international agencies and their efforts to cultivate an affective response that they deem reasonable. At the same time, these futures are put to work across domains of economic and political life. While it seems commonsensical to note that the present shapes the future, the present can also come to be told and worked on as an effect of the future (Nielsen 2014).

Nation-states have long tried to control environmental and natural resource futures, precisely because nation-states themselves seek to link past to future. Although the literature on nationalism has emphasized the importance of imagination (Anderson 1991), an often-overlooked point is that nationalisms imagine not only pasts and invented traditions, but also futures. The ideal futures of nationalism contain nations that will be properly constituted as a unified territorial unit, with people of like national status and language living in a form of social communion with each other. Nationalisms are future-orientated projects, then, a practice of doing political work in the present in the name of a yet-to-be-fulfilled future. The practices of environmental bureaucracies can also take on this future-orientated quality, as bureaucratic authority is often performed by translating celestial ideals into earthly practices, and by explaining the gap between ideals and reality as the result of human failings or mundane lack of resources (Herzfeld 1992; Mathews 2008). Such celestial ideals can be present-orientated, but they are often linked to ideas of improvement, towards a future that will be better than the present. What such studies reveal is that positioning by powerful actors in relation to the future is one of the ways in which they seek to perform their authority (Mathews 2014).

Another important future-orientated set of discourses are those of development, which is linked to nineteenth-century concepts of the emergence of immanent forces and energies, guided by officials or elites (Cowen & Shenton 1995). State-sponsored projects of modernization often classify people and places in terms of who or what belongs to the past or to the future (Fabian 1983) – classifications that often have a strong environmental component. From the early development theorists who saw societal progress as a temporal transition from the 'traditional' to the 'modern' (Rostow 1960) to recent efforts at crafting futures through quantitative benchmarks, such as the Millennium and Sustainable Development Goals, the project of development is founded on an imagination about what the future should be.

Visions of environmental futures have also had a powerful effect on environmental policies. Often, the environment has been understood by scientists and state environmental institutions in ways that link some parts of the landscape to particular temporalities. For example, forest services have commonly seen the most undisturbed areas of forest, such as mature high forest, as remnants of a formerly uniform forest cover (Cevasco & Moreno 2013; Fairhead & Leach 2000; Rackham 2006), driving officials to seek to restore landscapes to a desirable, densely forested future. Ecological theories of succession, influential especially in the first half of the twentieth century (Worster 1990), have also assisted forest services in classifying certain landscapes as degraded, the blame for which is frequently assigned to peasants and indigenous peoples (e.g. Sivaramakrishnan 1994). In many instances, the temporal classification of a landscape, whether as ancient or as recently disturbed, has been tied to a temporal classification of a people, as in the case of indigenous forest peoples, who have been seen by outsiders as inhabiting another time (Brosius 1997). The temporal and spatial frame that is chosen to classify part of the landscape in turn helps set the stage for future-orientated projects of nature protection, natural resource management, or environmental restoration (Cronon 1992). Frequently, these temporal and spatial ordering practices have deep and enduring colonial roots (Orlove 1993).

Thus visions of different futures and the relationship between present and future are central to political practices both by environmental planners and states, and by counter-movements that might oppose their policies. A prediction may sustain the legitimacy of institutions or infrastructures, and call into being specific policies in the present, or it may call into being opposition which undermines policies. In the case of oil, for instance, both a future with the resource (Weszkalnys 2014) and a future without (Limbert 2010) have profound implications for the cultural and political present. Often, it is the future's partial or complete unknowability that becomes a resource for those who wish to claim political, technical, or financial resources. Futures, then, like pasts, are a place of powerful imaginations on the part of rulers and ordinary people alike. People everywhere seek to make use of these energies in order to establish their legitimacy and authority. One area in which the future is powerfully linked to the routines of daily life is through modernizing projects of infrastructural change in the present, with related material, aesthetic, sensorial, and affective resonances (Larkin 2013).

Underpinning work on the political implications of environmental futures in the present are particular understandings of the state, whether explicitly articulated or implicitly held.[3] Vast as the literature on the state is,[4] we could distinguish between Marxist and Weberian visions of the state as a more or less unified structure, and related ideologies of modernization, improvement and development, or environmental degradation (Sivaramakrishnan 1998). James C. Scott (1998), for example, follows what might be called a neo-Weberian approach in tracing the impact of state ideologies of desirable landscapes upon the subjects of rule. Following more Foucauldian approaches, we can see visions of environmental pasts and futures as discursive formations that are internalized by the rulers and the subjects of rule, and used to justify colonial and postcolonial projects of appropriating natural resources (Fairhead & Leach 2000; Fairhead, Leach & Scoones 2012). Alternatively, following from science and technology studies and related approaches, we can see environmental futures as being produced through technopolitical work (Latour 2004; Latour & Woolgar 1987; Mitchell 2002). The authors of this issue draw variously on these approaches. Collectively, however, they understand the state as being fragile, enacted, and in motion. The state's ability to take a particular future and use it for political ends, therefore, becomes a complicated question. Indeed, like other official forms of knowledge, official and corporate practices of modelling futures seem not to silence opposition, but to produce new and unexpected results, sometimes increasing state control, but in other cases reducing it.

Imagination, also, plays a part in how different people respond to environmental futures. For scholars of science and technology studies, imagination and imaginaries, used more or less interchangeably, have the constitutive power to make facts through material and semiotic relations (Jasanoff & Kim 2009). This makes prognostic politics of interest not only for the more obvious political and economic effects that they might produce, of a dam or a road being built, but also for their power to constitute a particular truth, such as the presence of a mineral deposit. There is, however, a risk that imagination can come to stand for all that holds a cultural field together (Sneath, Holbraad & Pedersen 2009). Instead, we prefer to consider the cognitive and epistemic effects of imagination. Imagined national futures can help stabilize infrastructures in the present, or they can be used to stabilize particular facts about the world.

Circulations and transformations

As particular kinds of knowledge about the future move through everyday sites, their meanings and associations shift through diverse processes of translation (Rudiak-Gould 2012; West 2005). Understanding how people do political and technical work by invoking environmental futures requires particular attention to the question of knowledge transmission and transformation. How does knowledge about these futures move from those producing this knowledge to others? How does this knowledge morph along the way, as it is reworked by different groups in new locations? How is it selectively taken up in contrasting ways in different sites for particular political purposes? From the perspective of those who produce a forecast, this process of dissemination, through which that piece of knowledge about the future travels through different social spaces, opening up possibilities for 'resignification, semantic drift, miscommunication, and trouble' (Taddei 2013: 245), may be a cause of considerable anxiety.

There are different channels through which a prediction may circulate. Some channels may be 'official', such as company press releases or peer-reviewed journal articles. The government-authorized textbooks, national historiographies, and museum exhibits that Mandana Limbert analyses in this volume, for instance, tell a particular story about the past and future of oil in Oman. Others may be less 'official'. Maarten Onneweer (2014) describes, for example, the circulation of rumours in the Kitui district of Kenya about the presence of a resource of questionable existence – red mercury – and the potential profits that resource could bring. Nusrat Chowdhury also draws attention to more informal channels for the transmission of knowledge about the future. In her paper, the future of the Phulbari district in Bangladesh and the coal that lies beneath it circulates through different signs, from activists' graffiti on a wall to a multinational corporation's New Year's greeting card. Key to the circulation of a prediction is the question of trust. Continual circulation of this knowledge is contingent on those responsible for translating and passing on this knowledge considering it to be valid. To some, the trustworthiness may lie in quantitative or statistical parameters (Porter 1995); to others, it could be linked to the source of transmission (Lomnitz 1995). Indeed, official statistics may be seen as credible by some, the opposite by others (Mathews 2008).

Whatever the origin of a particular piece of knowledge about the future, as that knowledge moves through different sites, it is transformed. Gatekeepers to the circulatory process may emphasize some results over others. In her paper, for example, Jessica Barnes describes how climatic and hydraulic models produce projections of future rainfall and Nile flows that are highly uncertain, but which Egyptian scientists then use to predict an overall decline in the river flow. Such an interpretation may be linked to these scientists' personal and institutional location as well as their awareness of Egypt's vulnerability to declining water availability. At science/policy interfaces, there is often feedback between politicians and their modeller advisers and also a significant effect from modellers' own assessments of their political contexts. Modellers may, for example, express a higher degree of certainty about their predictions when communicating results to policy-makers (Lahsen 2009; Lövbrand 2007; Shackley, Risbey, Stone & Wynne 1999; van der Sluijs *et al.* 1998); in other circumstances, they may emphasize the uncertainty (Brysse, Oreskes, O'Reilly & Oppenheimer 2013; Shackley & Wynne 1996). At the interface between scientists and citizens, too, we see a process of challenge and negotiation, which may shift the nature and terms of the predictive science (Hébert, this volume).

Finally, just as important as thinking about the political work that an environmental prediction does in the present is to think about the political work associated with the *absence* of such a prediction. What happens when there is no prediction, or when the prediction is wrong? These are questions raised in the paper by Austin Zeiderman. He looks at the past failure of the Colombian state to predict a flood in Bogotá and at local residents' efforts to call the state to account for this failure. Avoiding the simple point that prognoses of the future are political when masquerading as technical, Zeiderman shows us how and for whom they become political. What matters is the particular manner in which the future is called upon. The future, then, is a realm of political and technical creativity open not only to states or corporations, but also to ordinary citizens. Here, we move fully from the vision of the future as the domain of state ideology or discourse to the future as something which people can make and do, with different futures proliferating and pulling at each other.

Papers in this volume

The papers in this collection offer ethnographic explorations of some of the new ways of using and addressing the future that are emerging in environmental domains. Some of the papers focus on resources, including water, copper, gold, oil, and coal – resources whose futures are shaped by a range of drivers, from global climate change to technological developments and evaluations of risk. Other papers focus on environmental resources' converse – environmental disasters, such as flooding or rapid sea-level rise.

The first papers in the volume look at ways of knowing the future. In Jessica O'Reilly's paper, she focuses on the future of the West Antarctic Ice Sheet under climate change – a future that holds particular significance owing to the impact of ice melting on global sea levels. O'Reilly takes us into the day-to-day practices of field science, modelling, and expert elicitation panels through which knowledge about the ice sheet's future emerges. Expertise about the ice sheet and predictions about the likelihood of its disintegration come not only from scientific data and technical knowledge, O'Reilly shows, but also through scientists' embodied experience, sensory engagements with the ice, and informal interactions with one another. Such ways of knowing, she argues, shape the production of knowledge about the future and help fill in the gaps of things that are fundamentally unknowable.

The second paper, by Jessica Barnes, looks at the climatic and hydrological models that scientists use to project the impact of climate change on future Nile flows. These models produce a range of results, which span from the positive – an outlook of increased water supply for Egypt – to the negative – a reduced water supply. Barnes examines how Egyptian and non-Egyptian scientists' presentations of uncertainty about Nile futures and their ability to know, address, and reduce it are tied to their positions in expert networks, particular material things, and the broader political context in which they are working. The paper underscores the importance of probing how scientists in different places – not only in Egypt, but in any political setting – interpret, deal with, respond to, and represent the uncertainty inherent in environmental events that have yet to take place.

The third paper, by David Kneas, looks at a junior mining company that explored for copper in the Intag region of Ecuador in the 2000s. Kneas examines the shifting registers this company employed to talk about the environmental future of this region. He describes, first, the story that the company told to investors on the Toronto Stock Exchange – a story of abundant copper and mineral wealth, which was founded on the use of particular sampling techniques, computer models, and geological theories, and the company's need to classify the deposit in terms that met the stock exchange's rules. He contrasts this imagined future with the future that the company told local communities – a future associated not so much with the profits of mineral wealth as with the removal of another threat: environmental degradation associated with agriculture.

Nusrat Chowdhury's paper draws attention to the role of different kinds of signs in signalling the presence of a resource and a potential environmental future. Her paper looks at a proposed open-pit coal mine in the Phulbari district of northern Bangladesh. This resource was framed by its proponents in terms of potentiality; a projection of prosperity into the future that tied its coal mines inextricably to the prosperity of the nation. Chowdhury highlights, in particular, the question of visibility, or, in the case of Phulbari, a lack thereof. The inability to see coal, she argues, demanded evidence in visual form – both for those who wanted its immediate extraction and for those against it. She documents how these actors collectively produced and consumed signs of coal's presence as well as the potential havoc its extraction might wreak for the thousands living above ground.

The second set of papers looks in more detail at the impact that various visions of the future of an environment or resource have on contemporary social worlds. Karen Hébert's paper looks at a proposed gold, copper, and molybdenum mine in the Bristol Bay region of southwest Alaska, which, if built, would be among the largest in North America. Hébert explores the political negotiations around a set of scientific predictions regarding the future of these resources and the risks associated with their extraction. She focuses, in particular, on a scientific report commissioned by the US Environmental Protection Agency. She examines the boundaries this report drew around what environmental futures were of concern, whose knowledge counted, and which timeframes were relevant. She argues that the tensions, or 'overflows' (Callon, Lascoumes & Barthe 2011), around the report reflect its overarching, if often frustrated, effort to separate scientific and technical truths from political contestations. While public participation in scientific assessments of risky futures can reinscribe exclusionary forms of knowledge, Hébert shows how it nevertheless opens new spaces for the exercise of politics.

Gisa Weszkalnys's paper focuses on São Tomé and Príncipe, a country where the existence of possible offshore oil has been identified, but the resource has yet to be commercially extracted. She introduces the idea of 'resource affect' to refer to the diverse affective resonances, including euphoria, excitement, scepticism, and trepidation, that emerge in contexts of resource prospecting, exploration, and extraction. She explores the nuanced affective horizons generated by the potential of an oil-based future in São Tomé. She looks, also, at the problematization of resource affect and the widespread efforts by international agencies to manage oil-related expectations and channel hope in São Tomé. Weszkalnys sees these efforts to curtail supposedly excessive expectations as an attempt to cultivate 'appropriate' affect.

In Mandana Limbert's paper, on Oman, she contrasts oil and another resource, water. Drawing on historical sources and oral histories, Limbert shows how the discovery of oil barely features in historical accounts of Oman, whereas water appears as a harbinger of promising futures. The future of oil and its depletion, on the other hand, is a constant refrain within political and economic discussions. Limbert's paper shows how some historical moments, activities, or transformations associated with natural resources become events, while others do not, and how some of these events are anticipated, but others are not. Here, again, we see how the time and place in which particular futures emerge can have long-term impacts, with some futures persistently recalled over an extended period of time, and other futures remaining relatively obscured.

The last empirical paper, by Austin Zeiderman, shifts the focus to look at conflicts over the failure of state science to know and govern the future not in the present, but in the past. Zeiderman draws a comparison between the flooding that affected Bogotá, Colombia, in 2011 and 2012, and the earthquake that hit the Italian city of L'Aquila in 2009. In both cases, people critiqued the state for failing to anticipate and prevent environmental disasters that took place. He argues that these cases reflect the constitutive relationship between political authority and foresight. While prognosis is central to political authority and legitimacy, Zeiderman shows how it may also be the ground upon which people make political claims on and critiques of the state.

The final commentary, by Elizabeth Ferry, draws out themes of temporality and uncertainty, anticipatory knowledge, and material signs from across the papers. Ferry reflects on the value of anthropological methods and approaches to cross-disciplinary discussions around the politics of environmental futures. The papers of this volume, she suggests, exemplify this anthropological strength, with their ethnographic depth and holistic perspectives.

Conclusion

The political impact of practices of imagining, modelling, and acting upon environmental futures is everywhere a feature of contemporary life, as competition for resources becomes more acute, and human domination of biogeochemical cycles intensifies (a process that some would argue marks a new era, the anthropocene

[Crutzen 2002]). Environmental futures are being envisaged in an increasing number of places, including through the scenario-planning practices of governments, scientists, and investors. These new ways of predicting or imagining the future build upon and complement existing regimes of calculation and prediction. We argue that contests over these prognoses are giving rise to new forms of nature, framings of time and space, and modes of politics.

Future-making often requires modellers to work with particular material objects and landscapes, from ice sheets to rock samples. These material entities are not easily subdued by prognostic practices. Rather, such practices always encounter limitations, making each future form potentially unstable, open to remaking, unmaking, or reinterpretation, perhaps by the politicians who commission them, or perhaps by the publics who are asked to believe them. It is in this messy process of negotiation that new forms of nature emerge. The sea off the coast of São Tomé becomes not just an expanse of water but a rock formation beneath the seabed that potentially contains a reservoir of petroleum. A river in Bogotá becomes not just a source of water but a transmitter and spiller of floodwaters. A stand of trees becomes not just a supply of wood and habitat for animals but a sink of carbon (Mathews 2015).

The process of looking to the future and making and remaking nature reworks our temporal and spatial categories. When modelling climate change impacts, the future becomes seen in blocks of decades or centuries rather than in terms of next month or next year. When a rock becomes seen as a resource – as in the silver of central Mexico that Elizabeth Ferry (2008) writes about – it produces a temporality that is linear, continuous, measurable, and finite. These shifting temporalities are closely tied to shifting spatialities. Imagining future climate change allows us to reimagine the global and the local, as well as the natural and cultural (Hulme 2010). When an ice sheet in one particular place – West Antarctica – starts to melt, it becomes tied to a global process of sea-level rise. To comprehend the future of Egypt's water supply, our gaze must shift to the mountains, thousands of kilometres away, where the rain falls that feeds the Nile.

Futures are not easily made, nor are people easily persuaded to believe in them. As both material and imaginative objects, always in motion, futures are open to challenge. The emergence of new ways of predicting the future has therefore produced new groups of actors who as collectives and individuals resist, modify, or make use of new kinds of environmental futures. These are emergent forms of politics, which do not look political in the classical anthropological sense (Dove 2000). The ethnographic settings that the authors in this volume write about press us to think about how new forms of politics come into being. This does not mean that we should discard our traditional disciplinary concerns with concepts such as citizenship and the relationship between states and subjects; rather, it shows us how a close study of practices of making the future can also make visible the emergence of new political collectivities. It would be easy, thinking with the categories of technoscience, to assume that there is only one world of future-making here, but the constitutive power of the imagination in practices of knowing the present and modelling the future means that multiple ontologies are potentially present, partially sensed, and bought into being by modelling practices. There is not one future, or even one kind of future that may be variously claimed (see Ferry, this volume), but multiple kinds of futures and of future-orientated politics.

Beyond looking at what our anthropological engagements with environmental futures can tell us about environmental politics and future-making, we are interested, also, in what these engagements might tell us about anthropology. Anthropologists have come to understand that we cannot seal ourselves off from our informants and their ideas, and that we have to be open to the possibility that their categories and cosmologies might trouble our own ways of thinking. As Marilyn Strathern's classical study demonstrated, the kinship relations of our informants can lead us to rethink our own understandings of gender relations (Strathern 1988) or the networks of technoscience (Strathern 1996). It seems fitting then that in writing about environmental futures, we take seriously the power of those who craft these futures to trouble the empirical practices and theoretical assumptions of anthropologists, as people who engage in future-orientated practices of our own. The scientists, officials, and ordinary people described by our contributors make the future tangible, in some degree addressable, through mundane practices of running computer models or classifying landscapes, just as anthropologists interpolate the future of anthropology through our mundane practices of citation and theory-building.

Looking to these parallels can give us pause for thought. The practice of writing literature reviews like this one, for example, and the tendency in doing so to evacuate matter and context from theoretical arguments, can too easily resemble the simple summaries given by climate modellers who omit the histories and contexts that give their models purchase upon the world (Lahsen 2005). In this introduction, therefore, our goal is to situate the papers of the volume, seeing each one as doing material, political, and theoretical work in the world, through practices of linking particular landscapes, people, and histories. For each author, the environmental futures at stake emerge from his or her ethnographic fieldsites. Rather than advancing theoretical points in the abstract, in these papers theory emerges in relation to particular research contexts, which it only imperfectly clarifies. The authors who focus on the more technical end of future-making (Barnes, O'Reilly, and Kneas) have paid more attention to the material and infrastructural aspects of modelling. Those who focus on nation-states and politics (Limbert, Weszkalnys, and Chowdhury), on the other hand, have drawn more upon the production and reception of national narratives of resource production. Finally, those who focus on science and risk (Zeiderman and Hébert) have tapped into work on the interface between experts and different publics. The places in which these authors locate their work make all the difference to their choice of theories, and to the way in which their theoretical frameworks and empirical material emerge in relation to each other.

Further questions are raised for anthropologists by ethnographically engaging with those who model, in a broad sense of the word, the future. Most importantly, studying different practices of crafting futures teaches us that empirical evidence and the linguistic predictions of models emerge in relation to each other, as successive models often call for a reclassification of data, of what matter is measured and how. Like modellers, we sometimes allow ourselves to forget that the material and the linguistic pull at each other through a process of ever denser co-mergence (Ingold 2012). Like modellers, we often allow ourselves to forget the histories of fieldwork which give rise to our data and sustain our theories. And like modellers, we can easily allow ourselves to make theoretical claims that are relatively divorced from our fieldwork and the limits of our evidence when we become entangled in professional or policy worlds. Studying modellers thus encourages us to change our understandings of the relationship between the material, the ideational, and the linguistic. Modellers describe tuning and parametrization as a skilled practice, which requires them to attend to the internal qualities of the model and to respond to pressures from their professional colleagues who are witnesses to their modelling practices. In addition, however, modellers feel constrained by their sense of the material world, often gained either in earlier experience of fieldwork, or through their daily life. We too, as anthropologists, need to cultivate our respect for the ways in which the material world presses itself upon our imaginations, even as our linguistic terms and categories (like parameters) fail to capture fully the unruliness of our human and nonhuman interlocutors. Matter becomes insistently political in environmental and natural resource futures, and we cannot, in advance, tell what kinds of politics will ensue.

NOTES

[1] There is also a rich body of scholarship on futures within a number of other disciplines, including science and technology studies (e.g. Brown, Rappert & Webber 2000; Fortun 2001) and sociology (e.g. Adam & Groves 2007).

[2] The literature on risk has become too large to summarize in a single essay or book, but see, for example, Power (2004).

[3] Visions of the environmental future in non-state societies, on the other hand, have a rather different quality, including no necessary assumptions of environmental improvement or degradation (e.g. Evans-Pritchard 1940).

[4] See Nugent (2010) and Sharma & Gupta (2006) for helpful syntheses.

REFERENCES

ABRAM, S. & G. WESZKALNYS 2013. *Elusive promises: planning in the contemporary world.* New York: Berghahn.

ADAM, B. & C. GROVES 2007. *Future matters: action, knowledge, ethics.* Leiden: Brill.

ADAMS, V., M. MURPHY & A. CLARKE 2009. Anticipation: technoscience, life, affect, temporality. *Subjectivity* **28**, 246-65.

ANDERSON, B.R. 1991. *Imagined communities: reflections on the origin and spread of nationalism.* New York: Verso.

APPADURAI, A. 2013. *The future as cultural fact: essays on the global condition.* New York: Verso.

BEAR, L. 2014. Doubt, conflict, mediation: the anthropology of modern time. *Journal of the Royal Anthropological Institute* (N.S.) **20**, 3-30.

BECK, U. 1992. *Risk society: towards a new modernity* (trans. M. Ritter). London: Sage.

BOND, D. 2013. Governing disaster: the political life of the environment during the BP oil spill. *Cultural Anthropology* **28**, 694-715.

BROSIUS, J.P. 1997. Endangered forest, endangered people: environmentalist representations of indigenous knowledge. *Human Ecology* **25**, 47-69.

BROWN, N., B. RAPPERT & A. WEBBER 2000. *Contested futures: a sociology of prospective techno-science.* Burlington, Vt: Ashgate.

BRYSSE, K., N. ORESKES, J. O'REILLY & M. OPPENHEIMER 2013. Climate change prediction: erring on the side of least drama? *Global Environmental Change* **23**, 327-37.

CALLON, M., P. LASCOUMES & Y. BARTHE 2011. *Acting in an uncertain world: an essay on technical democracy.* Cambridge, Mass.: MIT Press.

CEVASCO, R. & D. MORENO 2013. Rural landscapes: the historical roots of biodiversity. In *Italian historical rural landscapes* (ed.) M. Agnoletti, 141-52. Houten: Springer Netherlands.

COOPER, M. 2006. Pre-empting emergence: the biological turn in the War on Terror. *Theory, Culture & Society* **23**: 4, 113-35.

——— 2010. Turbulent worlds: financial markets and environmental crisis. *Theory, Culture & Society* **27**: 2-3, 167-90.

COWEN, M. & R. SHENTON 1995. The invention of development. In *Power of development* (ed.) J. Crush, 27-43. London: Routledge.

CRONON, W. 1992. A place for stories: nature, history and narrative. *Journal of American History* **78**, 1347-76.

CRUTZEN, P. 2002. Geology of mankind. *Nature* **415**: 3, 23.

DE LA CADENA, M. 2010. Indigenous cosmopolitics in the Andes: conceptual reflections beyond 'politics'. *Cultural Anthropology* **25**, 334-70.

DESROSIÈRES, A. 1991. How to make things which hold together: social science, statistics and the state. In *Discourses on society: the shaping of the social science disciplines* (eds) P. Wagner, B. Wittrock & R. Whitley, 195-218. Dordrecht: Kluwer.

DOVE, M.R. 2000. Bitter shade: throwing light on politics and ecology in contemporary Pakistan. *Human Organization* 63, 229-41.

EDWARDS, P.N. 2006. Meteorology as infrastructural globalism. *Osiris* 21, 229-50.

——— 2010. *A vast machine: computer models, climate data, and the politics of global warming.* Cambridge, Mass.: MIT Press.

EVANS-PRITCHARD, E.E. 1940. *The Nuer: a description of the modes of livelihood and political institutions of a Nilotic people.* New York: Oxford University Press.

——— 1963 [1937]. *Witchcraft, oracles and magic among the Azande.* Oxford: Clarendon Press.

FABIAN, J. 1983. *Time and the other: how anthropology makes its object.* New York: Columbia University Press.

FAIRHEAD, J. & M. LEACH 2000. Fashioned forest pasts, occluded histories? International environmental analysis in West African locales. *Development and Change* 31, 35-9.

———, ——— & I. SCOONES 2012. Green grabbing: a new appropriation of nature? *Journal of Peasant Studies* 39, 237-61.

FERRY, E. 2008. Rocks of ages: temporal trajectories of Mexican mined substances. In *Timely assets: the politics of resources and their temporalities* (eds) E. Ferry & M. Limbert, 51-74. Santa Fe, N.M.: School for Advanced Research Press.

——— & M. LIMBERT 2008. *Timely assets: the politics of resources and their temporalities.* Santa Fe, N.M.: School for Advanced Research Press.

FORTUN, M. 2001. Mediated speculations in the genomic futures markets. *New Genetics and Society* 20, 139-56.

GUILLEMOT, H. 2010. Connections between simulations and observation in climate computer modeling: scientists' practices and 'bottom-up epistemology' lessons. *Studies in History and Philosophy of Science Part B: Studies in History and Philosophy of Modern Physics* 41, 242-52.

GUYER, J. 2007. Prophecy and the near future: thoughts on macroeconomic, Evangelical, and punctuated time. *American Ethnologist* 34, 409-21.

HACKING, I. 1990. *The taming of chance.* Cambridge: University Press.

HASTRUP, K. 2013. Anticipating nature: the productive uncertainty of climate models. In *The social life of climate models: anticipating nature* (eds) K. Hastrup & M. Skrydstrup, 1-29. New York: Routledge.

HERODOTUS 1987. *The history* (trans. D. Greene). Chicago: University Press.

HERZFELD, M. 1992. *The social production of indifference: exploring the symbolic roots of Western bureaucracy.* New York: Berg.

HOLBRAAD, M. 2012. *Truth in motion: the recursive anthropology of Cuban divination.* Chicago: University Press.

——— & M.A. PEDERSEN 2013. *Times of security: ethnographies of fear, protest and the future.* London: Routledge.

HOWE, C. 2014. Anthropocenic ecoauthority: the winds of Oaxaca. *Anthropological Quarterly* 87, 381-404.

HULME, M. 2010. Cosmopolitan climates: hybridity, foresight and meaning. *Theory, Culture & Society* 27: 2-3, 267-76.

INGOLD, T. 2012. Toward an ecology of materials. *Annual Review of Anthropology* 41, 427-42.

JASANOFF, S. 1999. The songlines of risk. *Environmental Values* 8, 135-52.

——— 2004a. Heaven and earth: the politics of environmental images. In *Earthly politics: local and global in environmental governance* (eds) S. Jasanoff & M. Long-Martello, 31-54. Cambridge, Mass.: MIT Press.

——— (ed.) 2004b. *States of knowledge: the co-production of knowledge and social order.* London: Routledge.

——— 2005. *Designs on nature.* Princeton: University Press.

——— & S.-H. KIM 2009. Containing the atom: sociotechnical imaginaries and nuclear power in the United States and South Korea. *Minerva* 47, 119-46.

JOHNSON, A. & J. LENHARD 2011. Toward a new culture of prediction: computational modeling in the era of desktop computing. In *Science transformed? Debating claims of an epochal break* (eds) A. Nordmann, H. Radder & G. Schiemann, 189-99. Pittsburgh: University Press.

LAHSEN, M. 2005. Seductive simulations: uncertainty distribution around climate models. *Social Studies of Science* 35, 895-922.

——— 2009. A science-policy interface in the global south: the politics of carbon sinks and science in Brazil. *Climatic Change* 97, 339-72.

LAKOFF, A. 2008. The generic biothreat, or, how we became unprepared. *Cultural Anthropology* 23, 399-428.

LARKIN, B. 2013. The politics and poetics of infrastructure. *Annual Review of Anthropology* 42, 327-43.

LATOUR, B. 2004. Why has critique run out of steam? From matters of fact to matters of concern. *Critical Enquiry* **30**, 225-48.

———— & S. WOOLGAR 1987. *Science in action: how to follow scientists and engineers through society.* Cambridge, Mass.: Harvard University Press.

LILLA, M. 2008. The resistance of political theology. *Current History* **107**, 41-6.

LIMBERT, M. 2010. *In the time of oil: piety, memory and social life in an Omani town.* Palo Alto, Calif.: Stanford University Press.

LIVERMAN, D.M. 2009. Conventions of climate change: constructions of danger and the dispossession of the atmosphere. *Journal of Historical Geography* **35**, 279-96.

LOMNITZ, C. 1995. Ritual, rumour and corruption in the constitution of the polity in modern Mexico. *Journal of Latin American Anthropology* **1**, 20-47.

LÖVBRAND, E. 2007. Pure science or policy involvement? Ambiguous boundary-work for Swedish carbon cycle science. *Environmental Science & Policy* **10**, 39-47.

McCAMMACK, B. 2007. Hot damned America: Evangelicalism and the climate change policy debate. *American Quarterly* **59**, 645-68.

MacKENZIE, D. 2007. The material production of virtuality: innovation, cultural geography and facticity in derivatives markets. *Economy and Society* **36**, 355-76.

MASCO, J. 2010. Bad weather. *Social Studies of Science* **40**, 7-40.

———— 2013. Pre-empting biosecurity: futures, threats, fantasies. In *Bioinsecurities* (ed.) N.C. Sharp. Santa Fe, N.M.: School of Advanced Research Press.

MASON, A. 2007. The rise of consultant forecasting in liberalized natural gas markets. *Public Culture* **19**, 367-79.

MATHEWS, A.S. 2008. Statemaking, knowledge and ignorance: translation and concealment in Mexican forestry institutions. *American Anthropologist* **110**, 484-94.

———— 2011. *Instituting nature: authority, expertise and power in Mexican forests.* Cambridge, Mass.: MIT Press.

———— 2014. Scandals, audits and fictions: linking climate change to Mexican forests. *Social Studies of Science* **44**, 82-108.

———— 2015. Imagining forest futures and climate change: the Mexican state as insurance broker and story teller. In *Climate cultures: anthropological perspectives on climate change* (eds) J. Barnes & M. Dove, 199-220. New Haven: Yale University Press.

MAURER, B. 2002. Repressed futures: financial derivatives' theological unconscious. *Economy and Society* **31**, 15-36.

MITCHELL, T. 2002. *Rule of experts: Egypt, techno-politics and modernity.* Berkeley: University of California Press.

MIYAZAKI, H. 2006. Economy of dreams: hope in global capitalism and its critiques. *Cultural Anthropology* **21**, 147-72.

MUNN, N. 1992. The cultural anthropology of time: a critical essay. *Annual Review of Anthropology* **21**, 93-123.

NELSON, N., A. GELTZER & S. HILGARTNER 2008. Introduction: the anticipatory state: making policy-relevant knowledge about the future. *Science and Public Policy* **35**, 546-50.

NIELSEN, M. 2014. A wedge of time: futures in the present and presents without futures in Maputo, Mozambique. *Journal of the Royal Anthropological Institute* (N.S.) **20**, 166-82.

NUGENT, D. 2010. States, secrecy, subversives: APRA and political fantasy in mid-20th-century Peru. *American Ethnologist* **37**, 681-702.

ONNEWEER, M. 2014. Rumors of red mercury: histories of materiality and sociality in the resources of Kitui, Kenya. *Anthropological Quarterly* **87**, 93-118.

ORLOVE, B.S. 1993. Putting race in its place: order in colonial and postcolonial Peruvian geography. *Social Research* **60**, 301-36.

————, J. CHIANG & M. CANE 2002. Ethnoclimatology in the Andes. *American Scientist* **90**, 428-35.

PORTER, T. 1995. *Trust in numbers: the pursuit of objectivity in science and public life.* Princeton: University Press.

POWER, M. 2004. *The risk management of everything: rethinking the politics of uncertainty.* London: Demos.

———— 2007. *Organized uncertainty: designing a world of risk management.* Oxford: University Press.

RACKHAM, O. 2006. *Woodlands.* London: HarperCollins.

ROSENBERG, D. & S. HARDING 2005. *Histories of the future.* Durham, N.C.: Duke University Press.

ROSTOW, W. 1960. *The stages of economic growth: a non-Communist manifesto.* Cambridge: University Press.

RUDIAK-GOULD, P. 2012. Promiscuous corroboration and climate change translation: a case study from the Marshall Islands. *Global Environmental Change* **22**, 46-54.

Journal of the Royal Anthropological Institute (N.S.), 9-26
© Royal Anthropological Institute 2016

SAMIMIAN-DARASH, L. 2013. Governing future potential biothreats: toward an anthropology of uncertainty. *Current Anthropology* **54**, 1-22.

SCOTT, J.C. 1998. *Seeing like a state: how certain schemes to improve the human condition have failed*. New Haven: Yale University Press.

SHACKLEY, S., J. RISBEY, P. STONE & B. WYNNE 1999. Adjusting to policy expectations in climate change modeling. *Climatic Change* **43**, 413-54.

———— & B. WYNNE 1996. Representing uncertainty in global climate change science and policy. *Science, Technology, and Human Values* **21**, 275-302.

SHAPIN, S. & S. SCHAFFER 1985. *Leviathan and the air-pump: Hobbes, Boyle and the experimental life*. Princeton: University Press.

SHARMA, A. & A. GUPTA 2006. Introduction: rethinking theories of the state in an age of globalization. In *The anthropology of the state: a reader* (eds) A. Sharma & A. Gupta, 1-41. Oxford: Blackwell.

SIVARAMAKRISHNAN, K. 1994. *Modern forests: statemaking and environmental change in Colonial Eastern India*. Palo Alto, Calif.: Stanford University Press.

———— 1998. Modern forests: trees and development spaces in Southwest Bengal. In *The social life of trees: from symbols of regeneration to political artefacts* (ed.) L. Rival, 273-98. Oxford: Berg.

SNEATH, D., M. HOLBRAAD & M.A. PEDERSEN 2009. Technologies of the imagination: an introduction. *Ethnos* **74**, 5-30.

STRATHERN, M. 1988. *The gender of the gift: problems with women and problems with society in Melanesia*. Berkeley: University of California Press.

———— 1996. Cutting the network. *Journal of the Royal Anthropological Institute* (N.S.) **2**, 517-35.

SUNDBERG, M. 2009. The everyday world of simulation modeling: the development of parameterizations in meteorology. *Science, Technology, & Human Values* **34**, 162-81.

TADDEI, R. 2013. Anthropologies of the future: on the social performativity of (climate) forecasts. In *Environmental anthropology: future directions* (eds) H. Kopnina & E. Shoreman-Ouimet, 244-63. London: Routledge.

VAN DER SLUIJS, J., S. SHACKLEY & B. WYNNE 1998. Anchoring devices in science for policy. *Social Studies of Science* **28**, 291-323.

WALLMAN, S. 1992. *Contemporary futures: perspectives from social anthropology*. London: Routledge.

WEST, P. 2005. Translation, value, and space: theorizing an ethnographic and engaged environmental anthropology. *American Anthropologist* **107**, 632-42.

WESTHOEK, H.J., M. VAN DEN BERG & J.A. BAKKES 2006. Scenario development to explore the future of Europe's rural areas. *Agriculture, Ecosystems and Environment* **114**, 7-20.

WESZKALNYS, G. 2014. Anticipating oil: the temporal politics of a disaster yet to come. *Sociological Review* **62**, 211-35.

WORSTER, D. 1990. The ecology of order and chaos. *Environmental History Review* **14**, 156-70.

WYNNE, B. 1996. Misunderstood misunderstandings: social identities and public uptake of science. In *Misunderstanding science? The public reconstruction of science and technology* (eds) A. Irwin & B. Wynne, 19-46. Cambridge: University Press.

———— 2005. Reflexing complexity: post-genomic knowledge and reductionist returns in public science. *Theory, Culture & Society* **22**: 5, 67-94.

Pronostics : visions des futurs environnementaux

Résumé

Depuis toujours, l'esprit humain se livre à des pronostics. Les articles réunis dans ce dossier montrent que les nouveaux moyens de modéliser, planifier et interpoler l'avenir des ressources et des environnements se multiplient et occupent une place de plus en plus importante dans les politiques environnementales actuelles. Dans leur introduction, les auteurs évoquent deux dimensions de cette politique des pronostics : d'une part, les processus de formulation de prédictions pour l'avenir, et d'autre part les mouvements de ces prédictions à travers les institutions instables et désordonnées qui agissent sur le futur au présent. Ils font valoir que de nouveaux schémas de prévisions environnementales et de contestation de ces pronostics suscitent de nouvelles formes de nature, de nouveaux cadres de temps et d'espace et de nouveaux modes politiques.

Journal of the Royal Anthropological Institute (N.S.), 9-26
© Royal Anthropological Institute 2016

2

Sensing the ice: field science, models, and expert intimacy with knowledge

JESSICA O'REILLY *College of Saint Benedict and Saint John's University*

This paper analyses relationships that Antarctic glaciologists (including field scientists, modellers, and those working with remote satellite data) have formed with the Antarctic glacial environment. These relationships contribute to understandings about the tactile and experiential nature of scientific expertise, even when the experts claim to know little scientifically. Expert extrapolations about nature in the absence of data hint at the intimacy of field scientists with the environment with which they work, and of modellers with the virtual worlds with which they interact. However, my research on sensory engagement among scientists also makes apparent the ways in which the embodied and sensorial are not primitive, elemental, basic, or instinctual, but bound up in the complexities of nationalism, scientific translations of scale, and boundary skirmishes over what counts as expertise from within scientific disciplines. Expertise is formed in the spaces between intimate encounters in relationship with the weight of cultural learning that teaches experts-in-the-making how to encounter, analyse, compare, and interpret.

Before you go to Antarctica as part of a national research programme, you are outfitted with extreme cold weather (ECW) gear, along with layers that go underneath it. ECW clothing includes a big parka, down-stuffed snowpants, massive mittens that are attached to a cord that goes around your neck, balaclavas, sun goggles, and mukluks. The mukluks for Antarctica New Zealand (New Zealand's Antarctic programme) are blue, pillowy boots reminiscent of moon boots. They are not fit for skiing or walking since they lack support, but they are soft and warm, an Antarctic equivalent of slippers to wear around your campsite.

Camping on the Ross Ice Shelf as part of my ethnographic research on the relationships between environmental science and management in the Antarctic, I donned my mukluks every evening when we finished our observations and travels for the day. I was with eighteen other scientists and graduate students, most of whom shared my appreciation for the luxurious-feeling mukluks.

Our group was alarmed, however, when the ice shelf started melting underneath us. It was midsummer, and temperatures hovered around freezing, which made our ECW gear much too hot for most of everyday Antarctic camp life. A slight tick above zero degrees Celsius rendered our entire landscape soggy and heavy, with moisture seeping into our tents and our clothes. With the melting snow, our mukluks seemed to melt as well. As we padded around the trails we made between our tents, our wet footwear left blue imprints in the snow.

Some of my companions suggested scooping up the contaminated snow and taking it back to the base. I had done this the first morning, when I had tripped over a guyline and spilled my coffee. But as the steps proliferated, such a task seemed impractical. None the less, more than seven years after this Antarctic trip, I remember the blue footprints, the concern over leaving residue in the snow, and the seasonal warming that made the footprints visible and possible.

These tracks had nothing to do with the environmental monitoring research we conducted while in Antarctica, but were part of the phenomenon of field research in which we experienced the sensations of everyday life while conducting the formal rituals of scientific practice. The tracks, however, flagged one of the central concerns about the future of Antarctica: in a warming world, how will the massive Antarctic ice sheets behave? How much and how quickly can the melting ice sheets change global sea-level rise? While the footprints in the slushy top layer of snow indicated seasonal warming instead of a sustained global warming, they none the less inspired us to think about the insubstantial nature of the substance we were living on.

Although most of my research has focused on how Antarctic scientists produce knowledge through their research (O'Reilly 2011; forthcoming; O'Reilly, Oreskes & Oppenheimer 2012), in this paper, I analyse the other ways in which scientists come to know about the ice as they go about their work, both formally and informally sensing the ice. To do this, I read my ethnographic field and interview notes crosswise, for the moments in which scientists depict Antarctic ice casually and intimately, as people who know this substance in ways that most people do not.

This paper compares impressions from scientists relating to the experience of conducting Antarctic research. First, I describe field research in the Antarctic and the technoscientific encounters and misadventures my interview participants experienced. Antarctic terrain is sometimes difficult to work on, as scientists sample ice that melts, try to drill into unaccessed subglacial lakes, or simply must negotiate moving across a surface that shifts and melts. Along with participant observation over the past ten years in the Antarctic and in scientific laboratories where Antarctic data are analysed, I have interviewed fifty Antarctic scientists about their research. Interviews are open-ended and usually conducted after participant observation has been undertaken to enable our ability to discuss science and technology in depth. Next, I look at how data from the field become ice sheet models, as well as at some of the sensory and experiential impressions of human interactions with models. My last ethnographic section reflects on expert elicitation panels, a highly structured, subjective means for characterizing expert opinions on uncertain topics. Finally, the paper discusses how these experiences, subjective and personal, help to shape experts and the judgements they make about the future of the ice sheet. Relationships on the borders of discipline and expertise are mediated through sensory experiences with data, the alignment of Antarctic research with the nation, and the epistemic scale-making that grounds all scientific practice. These sensory engagements are not primary, primitive, or entirely outside of culture, but are trained, learned, disciplined, and contextual. This is where expertise is formed, in the spaces between intimate encounters in relationship with the weight of cultural

Journal of the Royal Anthropological Institute (N.S.), 27-45
© Royal Anthropological Institute 2016

learning that teaches experts-in-the-making how to encounter, analyse, compare, and interpret.

Sense, experience, and knowledge

Scientific ways of being in the world, though diverse and contingent on particular subjectivities, including disciplines themselves, are acculturated in particular ways. It is complex work to maintain ideals of sensing the world objectively, precisely, replicably, and truthfully. While technoscientific apparatus affords remote sensing, offering a perspective beyond that of simple human capabilities, scientific sensing takes place at all scales, including the intimate (Helmreich 2009). Much of the philosophy that helps us understand how scientists achieve this is viewed through the lens of epistemology (classic ethnographic examples include Gusterson 1996; Knorr Cetina 1999; Rabinow 1996; Traweek 1988). Studying how knowledge is produced approaches the intrinsic and extrinsic ways in which people work with the world, and with data, to make some coherent sense of them.

That individual scientists bring particular subjectivities to their research is well understood; indeed, particular subjectivities can be considered talent, a gift, or an ability to see nature in a novel way (see, e.g., Grandin & Johnson 2004; Haraway 1989; Ingold 2000). Evelyn Fox Keller titled her biography of Barbara McClintock *A feeling for the organism* (1983), referring to McClintock's often-repeated advice to elicit the truth from nature, even in the face of contradictory scientific knowledge. To Keller, McClintock's insistence on intimately knowing the plants she worked with – as individuals, not as specimen types – is a 'longing to embrace the world in its very being, through reason and beyond' (Keller 1983: 199). Keller suggests that McClintock's particular, groundbreaking scientific rupture was inspired through intimate knowledge of plants, knowledge so intimate that it was sublime, and subliminal. The particular tangibility of McClintock's research object – its materiality – was laid bare through the scientist's repetitive, individualistic, and consuming expertise.

Similarly, scientific knowledge can be formed through intimate interaction with the objects of study, in contrast to the ideal of a scientist as a detached, objective observer. For example, Carla Hustak and Natasha Myers analyse Charles Darwin's experiments with orchid reproduction as ones in which he 'participated actively with his experimental subjects, to such an extent that he *moved with and was moved by* them' (2012: 85, emphasis in the original). His experimental practices involved sensory engagement with his objects of study. His findings were only obtained through an encounter and relationship with the nature under scrutiny, instead of through remote observation.

Scientific understandings of nature rely on the global scale: making knowledge objective means making knowledge stable, repeatable, and ultimately scalable. Anna Tsing explains how contemporary environmental politics introduce processes of generalization whereby the global scale is merged with 'universal Nature' (2005: 90). Scientists tend to be disciplined into this idea: to be local is to be subjective or anecdotal. Climate science, with its prognostic hopes pinned to the global climate model, reifies the global scale into future imaginations. 'The global scale takes precedence', Tsing writes, 'because it is the scale of the model' (2005: 103). The technological tool meant to enable human thought about the future also sculpts the scale at which thinking takes place: the computer program shapes digital sensations of the future climate.

Sensory experiences of the environment are easy to find in the ethnographic record. Many studies focus on indigenous people's relationships with the environment, revealing, for instance, the complex knowledge of some indigenous understandings of ice across time and in local contexts (Carey 2010; Cruikshank 2005; Orlove 2005). These native knowledge systems are evidence-based, experienced-based, and imply a long-term intimacy with the known environment. In particular, Hugh Raffles analyses how 'local knowledge' is relational and intimate: that is, knowledge that is laden with 'affective sociality' (2002: 326). Rebecca Herzig (2005) developed her analysis of self-sacrifice – including self-experimentation – by scientists as a form of relating with the world, with truth, and with knowledge. Following this tradition, this paper examines how scientists walk the boundary between sensory, intimate experience and objective, universal knowledge as embodied people in the world. However, my research on sensory engagement among scientists also makes apparent the ways in which the embodied and sensorial are not primitive, elemental, basic, or instinctual, but bound up in the complexities of nationalism, scientific translations of scale, and boundary skirmishes over what counts as expertise from within scientific disciplines.

Sensory engagements involve linking up bodily experience with interpretative frameworks, broadly construed (Howes 2003; Stoller 1989). Sensing itself is not natural, but part of a cultural milieu in which people learn to take particular experiences seriously while discounting others (Geurts 2002). Knowledge is embedded, embodied, and placed, even when it purports to be universalistic, bureaucratic, or scientific: this knowledge is more than visual, it is sensed (Herzfeld 2005). This embodiment is taught, as in Natasha Myers's ethnography of a protein-folding class, in which the 'students must be willing to let molecular models instruct their bodies so that they can embody the fold' (2009: 188). Cultural framings of sensory engagements with the world are translated into knowledge, including forms of knowledge that comprise expertise. Among the multiple forms of expertise that exist in the world, expertise gained through sensory engagement can be considered 'interactional expertise' (Collins & Evans 2007). Sensory interactions are essential to cultivating expertise, though they are conventionally viewed as peripheral.

Below, I analyse some of the relationships that Antarctic glaciologists (including field scientists, modellers, and those working with remote satellite data) have formed with the Antarctic glacial environment, to contribute to understandings about the tactile and experiential nature of scientific expertise, even when the experts claim to know little scientifically. Expert extrapolations about nature in the absence of data shed light on the intimacy of field scientists with the environment with which they work, and the virtual future worlds with which modellers interact. This is why expertise and expert opinion matter. It is also why the physical, material, personal, and embodied characteristics of scientific expertise matter. To explore this area, I will consider the future of the West Antarctic Ice Sheet as the object of scientific concern that has inspired speculative, future-orientated research, and the implications of sensory experience for expert perceptions about what might happen to it in the coming centuries.

Doing science in Antarctica

In many ways, conducting Antarctic research is similar to doing the work of science anywhere, with some logistical adjustments made for the cold, wind, and isolation. However, the relationship between the nation and science here is explicit and direct. Notably, nations and their contractors run research stations and permit access to the bases and logistical support. There are sixty-eight national research stations in the

Antarctic run by thirty states.[1] The stations, owing to their origins as well as their current management, often assume a paramilitary atmosphere.

Stations are the logistical hubs for researchers. Support staff live there and help prepare research teams for field trips. While some researchers stay at the station, most venture out into the field to collect data, staying in tents or small structures helicoptered out. Some travel on 'deep field' expeditions which require flights or multiple days of land travel to reach, and are more logistically complex. Often, researchers pack up samples and specimens and transport them back to the research team's labs in their home country, though there are exceptions to this. Some samples and specimens may be analysed in field labs, which are commonly constructed from shipping containers and helicoptered in, or in more permanent lab facilities at the research stations themselves. Whether Antarctic materials travel back to 'the world'[2] or are analysed in Antarctica depends on many factors, including time and weather sensitivity, whether the researchers have travelled south by ship for an entire research season or are flown in for shorter durations, and the capabilities and equipment of the sponsoring national Antarctic programme.[3]

National Antarctic programmes administer the policies that govern Antarctic research. However, each national programme devises these policies through the international mechanism known as the Antarctic Treaty Consultative Meetings. These consensus-based, diplomatic meetings underscore the fact that while nations are responsible for the practices and behaviour of their scientists, they are in a nation-less space. The *laissez-faire* governance structure of the Antarctic Treaty System, with its emphasis on scientific research, international co-operation, and environmental protection, provides a framework that national programme managers interpret at their bases, with varying degrees of success.

Antarctic scientists conduct research in this remote place, under extreme environmental conditions and alongside more familiar political, bureaucratic, and material realities. While the mundane bureaucratic nature of much of Antarctic logistics does not often garner much attention from scientists, it none the less informs how they move about the place, conduct research, organize their camping practices, and survive. To confront the sensational Antarctic environment and landscape is to engage simultaneously with wildness and civilization. The expert knowledge that emerges below is a product of this engagement.

Crevasse

While the threat of ice melt and disintegration from climate change is often considered in terms of sea-level rise elsewhere on earth, field scientists manage the risk of the shifting Antarctic terrain in a more intimate sense. Most of Antarctica is covered in ice sheets, and therefore most human activity takes place on and around them. The ice sheet can be up to a mile thick, but it is not solid. It moves, shifts, and cracks.[4] Every person living in Antarctica receives mandatory training in camping, first aid, and emergency survival skills to cope with the cold, wind, and disorientating landscape. One of the many hazards to consider are crevasses, though people working in the Antarctic tend to be modest about the risks they face and play down some of the dangerous experiences they have. Bragging about one's near-fatal escapades is minimized, at least among one's Antarctic peers.

Crevasses are deep cracks in the ice sheet. These are often masked over by snow and are therefore difficult to detect in the white, monotonous landscape. Moving across the ice requires keen observation and, sometimes, climbing harnesses. Near tented field camps and bases you will find flagged routes which are areas of ice and snow that have already been surveyed as crevasse-free and are therefore simple to move across. But if you are off a flagged route, you must travel with a partner, be roped to that partner with a pair of climbing harnesses, and be ready to anchor that partner in case he or she, while leading, falls into a crevasse. Motorized vehicles, such as the tractor trains driving the route between McMurdo Station (US) and Amundsen Scott-South Pole Station (US), dynamited crevasses in front of their line of moving machinery to create a safe 'road' for transporting cargo. Landing aircraft on the ice sheet is particularly challenging since some crevasses can be difficult to visualize, especially when one is in motion.

In an interview with one glaciologist, Robert Lane,[5] he noted the difficulties of studying a location about which so very little was known, in rough terrain where one might unexpectedly have a closer than anticipated encounter with the ice sheet. Lane was a member of a long-term field study of the Siple Coast region of the West Antarctic Ice Sheet that involved a massive team of scientists and support staff flying out to a remote fieldsite each summer. Moving the drill and other major equipment to Antarctica involved over a year of planning, as the heavy pieces had to be shipped in containers from California.

Once the team and equipment arrived at McMurdo Station, they would load up a C-130 Hercules, a large workhorse military aeroplane capable of landing on ice. Pilots and researchers would fly over their proposed landing sites, taking aerial pictures to try to map out and avoid the crevasse patterns before landing. Once, however, the loaded-up C-130 missed imaging a crevasse and landed in one on Ice Stream D. Everyone on board survived and climbed out of the aeroplane, though the plane itself was severely damaged when the engines hit the ice.

This event triggered a massive recovery operation. Approximately thirty mechanics worked on fixing the plane for two months, winching it out of the crevasse, and then flew it out of the site. That season's drilling time was reduced from three months to one. The incident also changed the practice of landing C-130s on 'unimproved' snow and ice, reconfiguring how and with what equipment and field support glaciologists could work in the deep field. Now deep field projects are supported by the much smaller Twin Otter aircraft, and large stationary projects have come more into logistical favour. Since the extremely bulky hot-water drills can no longer be transported into deep-field locations, glaciologists have had to study the ice sheet with different techniques and equipment, as well as in new locations closer to base. This safety- and risk-orientated decision constrains research locations to those that are most practical instead of those most likely to yield significant data.

As in any field situation, the Antarctic environment makes particular research projects possible and impossible. Along with the complicated logistics operations that deep-field glaciologists have to contend with, the landscape which they are trying to survey is literally shifting, from stable to deadly, while the scientists are trying to track it. Crevasses dramatically remind one that the landscape is not a thin line on top of the earth; a landscape is multidimensional. The multidimensionality of the Antarctic ice sheet, particularly the marine-grounded ice sheet in the west, is tangible evidence that environmental shifts can take place in surprise bursts and in unanticipated directions. For scientists trying to predict the future of the ice sheet with limited data and models, anticipating the ice's surprising multidimensionality is a key problem. Learning this multidimensionality through life-threatening surprise is at once sensorial, epistemic, and eventually institutionalized.

Melting data

Though scientific knowledge is sometimes depicted as free-floating truth that a scientist manages to capture, scientists know that research objects are subject to environmental and human contingencies. Ice samples are precious, and to obtain them takes time, care, major logistical support through governmental, military, or contractor services, cost, and planning. Ice coring is a technical feat that involves drilling through hundreds of metres of ice in a way that cuts the ice sheet while keeping the ice intact. Once the ice is extracted it must be transported to a lab, either in the Antarctic or back in the researcher's home country, to be analysed. These labs contain precise sawing machinery, carefully managed temperature and humidity conditions, visualization equipment, and machines to analyse the small bubbles of air trapped in the ice. From ice cores, glaciologists can read the history of the ice sheet, including its age and behaviour, as well as the history of the earth's atmosphere, in the air bubbles trapped frozen in the ice.

When Marc Cash conducted his Ph.D. research in the 1980s, he made rudimentary observations at Ice Stream B, then took an ice core and prepared it for the National Science Foundation to ship back to his lab. To prepare the core for travel, he wrapped it in plastic, then in fur, and finally in a cardboard tube. That April, after the end of the Antarctic summer research season and upon the return of his ice core to his labs, Cash got a phone call around six in the morning that his shipment had arrived. In his characteristic colourful language, he recalled, 'I go screaming down to the office and I haul the ice off the freezer truck into [his adviser's] cold room and … I open it up and there's these little refrozen puddles of ice with fur in it. Melted down and they just refroze it and shipped it'. As he put it, 'My Ph.D. melted'. Cash and his adviser were able to mobilize their collegial networks to obtain ice cores in the next Antarctic seasons with support from the National Science Foundation, but the momentary panic of losing an entire season of research made an impression.

While Bruno Latour (1999) made painstaking efforts to describe the translations between lab and field, in Cash's case, the translations failed on a major scale, to the point where the nature-as-data became unintelligible. Here, melting ice represents not the fate of the planet, but one career trajectory and the challenges of reading nature from a distance, in laboratory settings instead of the field. The role of the nation, mediated through the National Science Foundation, explicitly winds through this tale as one of developing knowledge capacity for the nation through funding but also through enabling, complicated logistics. While the data read from the ice may be permanent, the material itself can be destroyed before scientists can 'read' it. Relying on the intactness of an almost intangible, ethereal, and certainty temperature-dependent substance lends a materiality to ice core research that is reliant on a carefully managed laboratory (and transport) microenvironment. Engagement with this material is full of care, enabled not only through precise scientific gestures but also through a governmental logistical system, which is bureaucratic and usually sensitive to the research materials at hand.

Penetration

Besides coring the ice sheet, scientists interested in what lies beneath the ice can drill down and observe the subglacial water systems, soils, and topography that sit on the Antarctic continental bed. There are several projects, historical and contemporary, that have aimed to penetrate the ice sheet to understand the physics, hydrology, and biology of the under-ice environment. These projects help develop theories about the future stability of the ice sheet in a warming world. For example, the penetration of subglacial Lake Vostok by the Russian Antarctic Programme emphasized a countdown involving metres and then centimetres until their drill penetrated through the ice into the lake. Delegates at the Antarctic Treaty Consultative Meetings expressed concern over the potential environmental repercussions of plunging into a lake with gallons of drill oil in the column above it. In response, Russia submitted successive, detailed environmental assessment reports to the Meetings over a period of years in order to gain consensual international support for penetration. The discourse around this project vacillated between narratives of discovering a new earthly frontier and properly mitigating risks of contaminating lake biota unknown to science.[6]

As governmental and public concern mounted over the oil-filled Lake Vostok drill hole, the technology of drilling and the ways that scientists approach drilling the ice sheet changed in the light of environmental scrutiny. Scientists now know that they are drilling not through pure ice into rock, but potentially into microbiospheres unique to the planet. To address this concern among scientific experts, the US National Research Council (2007) published a report on environmental stewardship of the Antarctic subglacial environment. In the 1990s, according to one glaciologist I interviewed,

> you could drop any kind of sensor, you could do tracer studies by pouring salt water or some dye tracer into bore holes, and nobody would blink. But it was actually discussion around exploring the large isolated subglacial lakes (or as they were thought about at the time as being isolated) that turned on the switch in terms of, well, if we want to study these environments, what about when we try to penetrate them?

The idea that one was sitting atop microbiological communities, and interconnected ones at that, proved instrumental for scientists in rethinking how they should approach their research, in terms of environmental ethics as well as acceptable scientific methodology.

Aside from the speculative possibilities of discovery, drilling in practice is mundane and routine. In an ethnographic case study, Martin Skrydstrup (2012) analyses the experimental anticipation of a drilling team on one of the domes of the Greenland Ice Sheet. He fuses the workaday experience of field camp life with the uneasy global scale of producing knowledge for climate models through the drill. At Skrydstrup's drilling camp, the cook makes fun of the scientists and their imprecision, everyone places bets on reaching bedrock, and the Primary Investigator (PI) pragmatically shifts her methodology to manage the risk of the bore hole collapsing. The people in this camp know ice, and it does not reduce neatly to either the sacred or the profane.

Scientific knowledge about the under-ice life of the ice sheet changed the way that scientists treat the ice sheet and do their work on it. The discovery of new life forms and flowing systems of water in the subglacial environment helped them reconsider the ice sheet as a habitat and as something influenced from below, not just on the exposed surfaces. In short, interaction with the ice sheet reconstituted how humans regard and approach the icy monolith, and the imaginative and scholarly possibilities inherent in it, including nationalistic posturing over knowledge frontiers and the discursive and material practices of discovery.

Journal of the Royal Anthropological Institute (N.S.), 27-45
© Royal Anthropological Institute 2016

Modelling

Researchers use ice sheet models to predict what will happen to the ice sheet in the coming centuries, using various future emissions scenarios. Since I had focused much of my early ethnographic work on Antarctic field sciences, I was surprised by the sensory appreciation that ice modellers showed, both towards their models and towards the physical ice. In an interview with one ice modeller, Trent Smith, I was struck by his motivation for Antarctic work: he did not do it to travel to and live in the Antarctic, but because of the mathematical and physics challenge of making ice sheet models. He summed it up by comparing modellers to glaciologists: 'Most glaciologists are climbers or outdoor people, and they like that, and then they get interested in what they're climbing over . . . but I don't want to go for a walk on it . . . that's not the motivation'. He was initially planning to study ecology, but after finding that he was spending all his time at university 'chopping up rats', he searched for something more compatible with his interests. He found it in an environmental science course, remembering one lecture in particular about energy transfer and solar radiation. 'They were showing equations . . . and something clicked', he recounted. 'I could see that the equations reflected something that you could observe and that was happening in reality, and that got me very interested in the use of models and maths to understand the environment'. Once Smith started applying for Ph.D. positions, he simply looked for modelling – modelling anything – in the title. His doctoral work involved modelling ice sheets, but he wanted a break from academia by the time he completed. Therefore, he began his professional career by working as an engineer modelling urban sewer flow in a government agency before eventually returning to ice sheet modelling.

Modellers joke about their nerdy reputation compared to the apparently more adventurous and athletic field scientists.[7] A modeller named Tim characterized the social caricatures as follows:

> The classic modellers, you know, I've got spectacles, for example . . . a typical modeller is introverted and not particularly practical . . . whereas the typical field worker is very much a can-do person and craves excitement. If you're an outdoor person and like climbing mountains, you've obviously got a slightly different outlook on what you want from life than a modeller, someone who enjoys solving differential equations.

He went on further to describe the power dynamics between the two groups, since modellers often consider the role of field scientists in the Antarctic to be to test the ground-truth of physical processes and collect data for the models. Field scientists, in contrast, are often looking not for basic data but for exceptional, dynamic events. Tim explained that 'modellers want field people to analyse something that's pretty predictable so that they can put it into their models, whereas field workers want to go and find something new and exciting and essentially unpredictable'. This puts the very act of scientific analysis and, even more basic than that, the very act of scientific observation at odds. The kind of looking one wants to do is related to disciplinary ideas about truth, discovery, and even what makes an appropriate disciplinary research narrative.

The delineation between field scientists and modellers is not always so stark, however. Marc Cash, whom we referred to above, a glaciologist who works in both modelling and field sciences, does not find the division as cut-and-dried as his peers above. This is similar to the porous though policed field-lab borderlands that Robert E. Kohler (2002) analyses in his history of biology. According to Cash,

> It's interesting, there are real field people who wouldn't crunch an equation and there are real modellers who don't do much in the field. But a whole lot of the people who are doing it are shaded somewhere in between. The . . . [names early glaciology programmes] tradition really was you go out and make the best observations and then you turn them into the best model.

Indeed, Cash's career spans both field science and modelling, to the point where he could not remember which team he played on in a soccer game between the modellers and the 'data' (though he does recall he scored a goal). His use of the word 'data' to stand in for field scientists, who gather data among other, more complex activities, is telling, though used in a joking manner.

How modellers interact with their models and data is well studied (Edwards 2010; Lahsen 2005; Oreskes, Shrader-Frechette & Belitz 1994). While this work analyses how models come together, my research focused on how models do not come together, as well as on how modellers pragmatically lived and worked with their always-partial computational representations. Ice sheet models turn the ice sheet into a series of equations, and modellers sometimes slip from talking about ice sheet behavior into mathematical language. For example, modeller James explained, while discussing the difference between ice sheets (which sit over land) and ice shelves (which float over water),

> They flow in a very different way, these two ice masses. Ice sheets flow by – because they have traction at the base so they in fact flow by what is called shearing, so they flow by layers of ice which shear on top of one another while an ice shelf is like you would drop oil on water. It spreads in all directions. It stretches and it spreads. It's floating on the seas. This is a totally different way of solving the force balance than solving for the flow. This leads to fundamentally different mathematical equations.

Here, James turns the ice into an equation to be solved. His role is to provide calculations to fill in the gaps where observations have not explained how glacial physics work: that is, modellers create approximations that keep the model running.

Harry, who is both a field observer and a modeller, is uneasy about these approximations in numerical modelling. He prefers working with what he calls simple models, which are not run into the future but instead can be tested for accuracy in real time. He told me,

> Numerical modelling is always kind of making the best of it and it's what I don't like about it. So I'm more like a slightly old-fashioned scientist and want to do simple models. I'm more analytical with mathematics so you know exactly what you're doing though you have your limitations.

Harry avoided the more speculative possibilities of future-orientated models and instead worked within the confines of simple models, to which improvements also flow into numerical models.

Phillip, a remote sensing specialist, also underscored the limitations of modelling, particularly ice modelling, which has a long way to go in terms of reflecting the glaciological processes occurring. As we sat in his office during an interview, he said, 'The ice sheet modelling community has been living in its sort of fake world for a long time . . . maybe modelling square ice sheets and square ice shelves because they didn't have much else to work with'. That the models turned the ice into 'squares' instead of the natural Antarctic topography illuminates the complexities of trying to create standard models from a haphazard, random, and contingent world. That the terrain cannot be depicted with realism raises the question of how the processes driving ice behavior are being simplified, computerized, and modelled. It also suggests that the picture of the future ice sheet being presented is likewise limited, partial, or even inaccurate.

Journal of the Royal Anthropological Institute (N.S.), 27-45
© Royal Anthropological Institute 2016

Modellers are practical about the mathematical, physical, and computational limits of their models as programs that can represent nature, with the obvious caveats about limitations and unknown and poorly understood processes. Modeller Tom described thinking about nature and limited knowledge in this example:

> *Tom*: The trick is that you don't know anything about the water, really. You just know that it's wet at the bottom [of the ice sheet]. And then there's a number which is called the sliding viscosity, which relates the sliding velocity to the forces acting. And then this sliding velocity, *you've hidden that, you don't know anything about the water* [emphasis added]. So it's possible from satellite data to work out what the sliding viscosity is, or make a reasonable estimate of it. But you don't know how it's going to change as the amount of water changes, so you either pretend that it's not going to change, or you use some kind of untested model to say it's going to change by such and such an amount if something happens upstream and there's more water coming down underneath the ice stream.
>
> *J. O'R*: So, like an educated guess.
>
> *Tom*: It probably would be nice if it was an educated guess.

In light of contrarian scrutiny, where such phrases might be held as 'proof' of a vast scientific climate conspiracy, it is important to think about the context that the modeller is referring to. In this case, the modeller is 'fudging', or making an approximation that connects chains of known calculations, backed up with observational data. In complex-systems modelling, the pieces of natural processes must fit together. In the absence of observations, something must stand in to make the entire model run. These 'known unknowns', where some equation stands in for a knowledge gap, create uncertainties in the model, but these can be described, quantified, and marked for future research. While Tom candidly describes an approximation, it must be noted that to know one is fudging an equation, and being open about it in one's methodology, is considered preferable by other scientists to unwittingly ignoring a process in the model entirely.

None the less, the interim calculation that modellers choose to approximate a 'known unknown' in their model is a guess based on immersion within the modelled ice sheet, an understanding that is expert as well as partial, and an attempt to work with their state of not knowing something along with what they know well (or at least as well as any other expert in their area). With approximations, modellers are working on the edges of what they know and what they do not. These edges are negotiated through their numerical, data-rich relationship to the model as well as their pragmatic understandings of the model's always-partial reflection of the world it attempts to signify.

Expert elicitation

In my ethnographic research with Antarctic and climate scientists, no scientific activity has been as internally controversial as 'expert elicitation', another technique for characterizing expert impressions of the unknown. Subjective and personal, it gives some scientists an impression of soft data, of social science. None the less, expert elicitation projects have done what models and semi-empirical methods (Pfeffer, Harper & O'Neel 2008; Rahmstorf 2007) have not yet been able to do: describe with some precision the likelihood of rapid disintegration of the West Antarctic Ice Sheet. This concern over the future of the ice, as well as its related global sea-level rise, has provoked several decades of glaciological research in the Antarctic. This future haunts everyday Antarctic research and logistics, in addition to the fate of the planet. Expert elicitation forms moments where scientists leave their practised, rational moorings and have to rely on sensory engagements, which are less explicit but still formed in the sociocultural milieu in which they live and work.

Since the 1950s, glaciologists have marked the West Antarctic Ice Sheet (WAIS) as a site of concern. This concern includes the geological instability of the ice sheet, since much of it sits below sea level and therefore interacts directly with the warming ocean. Scientists are also interested in the potential of metres of global sea-level rise that the ice sheet contains. In addition, the complex geophysics to explain ice sheet behaviour seem to stand just slightly outside of expert comprehension; there are still some mysteries to how the marine WAIS works at a fundamental level. For example, modelling difficulties have made it challenging to project the likelihood of a rapid disintegration of WAIS, which would have major impacts on global sea-level rise. Two recent publications (Joughin, Smith & Medley 2014; Rignot, Mouginot, Morlighem, Seroussi & Scheuchl 2014) are the first to suggest that the ice sheet has already passed 'the point of no return' at which disintegration has been put into irreversible motion. The future of the ice is difficult to predict and a major amount of human effort and resources is being poured into attempts to study it.

The genesis of expert elicitation about WAIS emerged in collaboration with David Vaughan from the British Antarctic Survey and John Spouge of Det Norsca Veritas, an insurance company specializing in risk assessment, as insurance companies do. The project was published in *Climatic Change*, a well-regarded climate science journal (Vaughan & Spouge 2002). The insurers went through several iterations of trying to gather enough information to perform an adequate risk assessment. First, they looked at the published literature, but there simply were not enough publications to be significant. Next, they tried to put together a statistical model – akin to using actuary charts for life insurance – but, similarly, there just were not enough numbers to make one that was credible. Finally, the group decided to use the Delphi approach, which has been described to me in an interview as a method 'where you ask people whose judgement you think is better than average'. Clearly, the decision about whom to ask, and what is considered judgement, is remarkably fraught and subjective in the most basic sense of the word. These selected experts gauged their sense of two scenarios, one high (20 cm sea-level rise/century over 200 years) and one low (5 cm sea-level rise/century over 200 years), as well as how certain they felt about each particular scenario occurring. In the end, virtually no scientist thought the ice sheet would collapse in 100 years, but most people thought it could go in 1,000 years. But in between that 900-year spread, the researchers found little consensus (Vaughan & Spouge 2002).

The Delphi method was developed by researchers in the RAND Corporation in the Cold War 1960s (see Dalkey 1967; Helmer 1967) and has been employed by social researchers since with various critiques and modifications. In its traditional form, researchers employing the Delphi method send out a series of surveys to an anonymous, selected group of experts. The first survey set is open-ended, with subsequent surveys using previous survey results to attempt to achieve consensus (including quantified uncertainty) over the survey series (Hsu & Sandford 2007). Expert elicitation methods, more broadly, draw from the Delphi approach but may also include in-person expert meetings in which assembled experts provide projections on a future-orientated matter and work through consensus-making techniques to try to provide a range of group projections.

These methods are used in attempts to answer technoscientific questions that quantitative data and scientific observations cannot. According to Delphi developer Olaf Helmer, expert elicitation methods allow social scientists to 'try to obtain the relevant intuitive insight of experts and then use their judgments as systematically as

Journal of the Royal Anthropological Institute (N.S.), 27-45
© Royal Anthropological Institute 2016

possible' (1967: 4). As the examples above suggest, however, the intuition being assessed is as culturally crafted and contingent as the general technical and knowledge skill sets conventionally associated with expertise. These methods allow for inclusion of sensory engagements and experiences with landscapes, environments, and objects beyond the traditional scope of scientific research. Expert elicitation strategies attempt to harvest the pieces of expertise – intuition, formed and trained – that lie outside the scope of the formal experience of scientific practice. In these techniques, experts are tasked with infusing their sense of the future with their personal experiences alongside their scientific findings.

Some people who work on WAIS have reservations about expert elicitation projects, and there appears to be a division between those who view these studies as reasonable stop-gap undertakings, and those who regard them as sloppy research incapable of saying anything credible. A modeller named Pierre was heavily critical of this project, asking me to turn my recorder back on after an interview to speak at length, on record, about his opinion. He called expert elicitation 'a poor man's solution if you don't have anything else. If you don't have good scientific arguments, ask our friends what they think!' To him, expert elicitation, with its subjective methodology and human vagaries, is a decidedly unscientific way of polling scientists about matters of their expertise. It plumbs the intimacy of expertise when scientists cannot use their conventional techniques to seek answers, suggesting that expert 'gut instinct' might have value beyond the general public's sensibilities.

The Intergovernmental Panel on Climate Change (IPCC) is a massive climate assessment that produces reports about every six years. The IPCC is jointly run by the United Nations and the World Meteorological Organization, and hundreds of scientists participate in writing and reviewing its reports, which focus on the physical science basis of climate change, impacts, adaptation, and vulnerabilities to climate change, and climate mitigation. In the 2013 IPCC assessment report (called AR5), the authors decided to include expert elicitation publications for review, despite their controversy. The authors may have been trying to represent more of the information available; the 2007 attempt at representing WAIS's potential for rapid disintegration was a decided punt (Oppenheimer, O'Neill, Webster & Agrawala 2007; O'Reilly 2015; O'Reilly *et al.* 2012).

For example, the expert elicitation research of Bamber and Aspinall (2013) received several mentions in the 2013 IPCC chapter on 'Sea Level Change'. The IPCC authors cite Bamber and Aspinall to underscore the poorly constrained probability of grounding-line[8] retreat contributing to global sea-level rise through 2100, uncertainty over whether recent changes in WAIS are the start of a long-term trend of disintegration or are simply regular variability (IPCC 2013: 1185), and the 'wide spread' of expert elicitation-based sea-level rise projections, which indicates 'a lack of consensus' on the probability of ice sheet collapse (2013: 1186). That the IPCC authors chose at all to include expert elicitation to a serious degree can be interpreted as a change in the types of information considered in the physical science-based Working Group I of the IPCC, but the interpretation of expert elicitation research in this case is used to underscore uncertainty and not knowing.

In Bamber and Aspinall's article (2013), though, the authors emphasized a quantifiable finding from their expert elicitation research. While noting that expert opinion is both 'very uncertain and undecided', they wrote that, 'we find that the median estimate of such contributions is 29 cm – substantially larger than in the [IPCC's Fourth

Assessment Report] AR4 – while the upper 95th percentile value is 84 cm, implying a conceivable risk of a sea-level rise of greater than a metre by 2100' (2013: 424). In the later published IPCC AR5, the IPCC provided model-based estimates of sea-level rise. Its results ranged from 28 to 98 cm, with a mean from 44 cm to 74 cm for various future climate scenarios. Furthermore, these numbers are only from the likely range (with a 66 per cent confidence interval), not the worst-case scenario. In other words, the models provide more dramatic sea-level rise projections than the experts do.

Some scientists express concern that expert elicitation provides subjective results and may skew towards the more dramatic side of potential futures. But this case, along with several others, underscores what I and others have termed 'erring on the side of least drama', in which experts under-predict or downplay results (Brysse, Oreskes, O'Reilly & Oppenheimer 2012; Hansen 2007). I suspect this may have to do with long-standing scientific cultural values of being 'modest', objective, and rational (Haraway 1997; Shapin & Schaffer 1989), in addition to being a response to contemporary climate contrarianism.

Additionally, the research community studying WAIS is small, and findings such as those from various model runs might be informally circulated among experts ahead of formal publications of research. Conversation among peer networks, presentations at conferences, and meetings and opportunities for peer review allow some research findings to gain traction ahead of publication. These conversations and informal circulations of research are part of the subjective and intimate work of experts.

Expert elicitations evoke and use non-scientific types of knowing among scientific experts, which Harry Collins and Robert Evans (2007) call 'specialist tacit knowledge', noting the multiple forms and dispositions of expertise that intersect, counter, and bolster such knowledge . Expertise and judgement relate, but not simplistically. Though the methodology of expert elicitation is easily opened up for critique, particularly by the experts who participate (or who choose not to participate), the idea that scientific experts may make expert judgements that are 'better than average' and rooted in embodied experience of the subject matter at hand lets us think about how expertise is formed, and how the formation of the expertise, not just the resulting data and analysis, may contribute to helping understand some of the unanswered or little-understood questions about nature and its potential future trajectories.

Knowing Antarctic futures

In his laboratory, Robert Lane concluded our interview by talking about climate change. He recalled talking with Al Gore in the cafeteria of McMurdo Station, telling him how Antarctica seemed to be changing in the decades he had spent as a researcher there. When he started going south, he had no idea about climate change, 'then suddenly in 1996, I saw water running through McMurdo, and then later I saw the dripping water into the shear margins. Then I went to the Trans-Antarctic Mountains and there was water pouring down the slopes into these outlet glaciers from East Antarctica'. Lane suddenly had a concrete impression, like my group and our bleeding mukluks, of the icy continent melting. The tactility of the melting, the living in it, dealing with equipment in wet places that are supposed to be frozen, makes an impression, even if the phenomenon is an isolated event not related to climate change. Like the news stories and extreme weather events in our home places, we cannot help but wonder if these things we have never seen before are part of a broader phenomenon.

Journal of the Royal Anthropological Institute (N.S.), 27-45
© Royal Anthropological Institute 2016

In this paper, I have analysed the ways in which glaciologists sense the ice, build predictions, and deploy anticipatory means for explaining their scientific research. I have discussed how scientists working in future time fill in gaps in their predictive capabilities with expertise of the ice sheet that is both intimate and technical. Scientific research is planned around the hope of witnessing and studying natural phenomena; futures are imagined through the conduct of research and the act of living in the Antarctic field. It must be noted that experts are doing the imagining in this case. How does expertise influence these visions of the future? In the examples above, expertise extends beyond epistemological matters when embodied, everyday experiences that accrue alongside the practice of scientific research are taken into account.

This paper has focused on how scientists' ways of being in the world envelop more than their disciplinary foci. Their narratives about these experiences can help us understand how sensory engagements are learned and integrated into expert scientific knowledge more broadly. Sensing ice sheets is a complex activity – these massive and chaotic formations have challenged scientists to find a vantage from which humans can attempt to understand them. For example, in one interview, a remote sensing specialist who also does primary field observations notes that the interior is a 'binary landscape, snow and sky and that's all you see'. But on the coast (of the Greenland Ice Sheet), 'you see the whole thing. You see calving glaciers. You see icebergs. You see a much more dynamic part of the ice sheet'. Glacial dynamism, such as collapse, underscores the potential of the ice sheet in spectacular fashion. Witnessing glacial activity at any scale involves situating oneself in relationship to the ice. There is a relational perspective to satellite observations as well, providing a God's-eye perspective of the continent and its subglacial dramas. This observational work, both professionally and as a consequence of living in the field, may serve as expertise, or bias, or anecdote, to the scientists when they are asked to project their sense of the ice sheet's future.

The observational data, then, must be incorporated into models to make them as realistic as possible. As described above, models are simply complex computer programs, coded to mimic the earth's systems. Since these systems are imperfectly understood, it is essential to underscore that climate models are simply approximations. At each junction where a system is poorly understood – how WAIS will melt, how the ocean will expand as it absorbs CO_2, or how the radiation from clouds protects or magnifies warming – the modellers write code that is an approximation of an approximation. These are mathematical fixes that allow the computer code to run, but without any numerical representation of nature whatsoever. But again, modellers cannot provide the high-resolution projections needed if there is high uncertainty about the behaviour of the ice sheet or a simple lack of data overall. Therefore, the models contain approximations even at a high resolution.

None the less, in response to several governmental and NGO inquiries about the future of WAIS, scientists make projections. How do they do this if the model codes contain huge runs of approximations? In this paper, I have suggested that they formally and informally assess the data at hand and relate that to their experiences with the ice sheet. This is where the intimacy of field studies comes in, where people live on the ice sheet, gather pieces of it to inspect, or even crash their planes into its crevasses.

These data, obtained through field studies or by flying high above the continent, and the research stories and histories that glaciologists form around them help people make projections that show up not in formal IPCC documents, but in lectures, presentations, and interviews. Through engagements with intimately, intricately extracted data, glaciologists form a sense of the ice that anticipates the behaviour of the future ice sheet.

Such anticipation plays out informally in the ethnographic examples described above, as glaciologists and modellers confront the limits of their technoscientific expertise. As also shown above, the field sciences are emplaced, logistics can fail spectacularly, and models need stopgap codes to keep them running. These professional crises are experiences that shape how scientists perceive and inhabit their world, including, I suggest, their interpretations of the future of the ice sheet.

This work is conducted formally in expert elicitation projects, in which research brings experience to bear, in an attempt to make predictions about a world not yet understood scientifically. Expert elicitation research affords an opportunity to quantify these felt, subjective, intuitive, and intimate experiences with the ice sheet and compare them with data obtained through standard process-based models. At least in the most recent case, the expert elicitation project and the model projections are noticeably similar, and the expert elicitation predictions are more conservative than the objective model data. In this case, experts use their technical as well as experiential skills to say something provisional about an unknown and uncertain future (see also Collins & Evans 2007). Though expert elicitation remains on the edges of acceptable scientific methods owing to its inherent subjectivity, it provides an opportunity to give policy-makers a little more analytical and predictive purchase for making climate decisions.

This paper shows how expert judgement contrasts with popular or public judgement, not only owing to epistemological factors but also through intimate engagement with the world that scientists research. This world is created through long-standing interaction between the researchers and the ice that they act upon – studying it, certainly, but also living upon, falling into, and feeling saturated by it. Though difficult to quantify, the experiential domain of scientific practice contributes to the patina of expertise along with training and skill. By interacting with the ice and their iterative depictions of it, experts manage to put together particularly informed visions of the future.

Interacting with the nature of one's scientific inquiry is a world-making project in addition to a knowledge-making one. To live in a glaciological world, in comparison to a world populated with glaciers, is to feel and read specific signs, to speak a particular, specialized language, and to forge comparisons with distant places and forms. In this case, science becomes not only a way of knowing, but also a way of being, an ontological effort that is relational and practical with regard to the objects of inquiry (Latour 2013). That is, while the disciplining nature of science may filter a scientist's interpretation of the world at hand, the world itself participates, by setting itself up in a way that makes some relational options easier or more logical and others virtually impossible (Kohn 2013). By examining the intersection between the experience and expertise of Antarctic glaciologists and modellers, I suggest that sensing the ice is not only an explicitly cultured and disciplined endeavour, but also a site where knowledge is formed through embodied experience.

NOTES

Parts of this study were conducted as a postdoctoral research associate, with the mentorship and support of Naomi Oreskes, Michael Oppenheimer, and Dale Jamieson. While my analysis and mistakes are my own, I thank them for their guidance along with my colleagues Keynyn Brysse, Matt Shindell, William Thomas, and Milena Wazeck. Many thanks to Jessica Barnes for her invitation to participate in this issue, and for her close and generous reading of my drafts. Thanks to colleagues who have provided insights to various iterations of this article, including Martha Lampland, Fabiana Li, Andrew Mathews, Jerome Whitington, and two anonymous reviewers for *JRAI*.

[1] The Antarctic Treaty (1959) suspends the territorial claims of seven claimant states: the United Kingdom, New Zealand, Australia, Argentina, Norway, France, and Chile. The Treaty also forbids any other states from making claims, an agreement sought owing to increasing interest in claims from the United States and the then Soviet Union. The nine states above plus Belgium, Japan, and South Africa are the original Treaty signatories. More states have been added; all consultative parties have a sustained and active scientific presence in the Antarctic. Other states without an Antarctic scientific programme can sign the Treaty and observe the annual Antarctic Treaty Consultative Meetings as non-consultative parties, along with a small number of expert and observer groups.

[2] Some Antarctic people distinguish Antarctica from what they call 'the world', which extends the scale of Antarctic discourse past the global to the apparently extraterrestrial.

[3] While this paper focuses on glaciologists and ice sheet modellers, both of whom are concerned with the future of Antarctic ice, Antarctic research interests are diverse. The South Pole is home to the world's largest neutrino detector, planted a kilometre down into the ice to try to capture extraterrestrial neutrinos as they zip through our planet. Biologists and ecologists examine non-native species, fungal communities, and the multiple species and colonies of penguins. The small area of the Antarctic that is ice-free is a focus both of intense scientific interest, owing to the remarkable natural processes and species found there, and of infrastructure development, as it is much easier to build on solid ground instead of snow. The rare and unique environments in Antarctica have inspired some of the world's most creative and comprehensive environmental management strategies.

[4] Ice sheet behaviour occurs independently of warming and other anthropogenic forcings. However, some parts of the Antarctic ice sheets are changing more rapidly owing to warming air and oceans.

[5] Interview participants' names are pseudonyms.

[6] You may consider consensus in the Antarctic Treaty System as earned when all questions have been addressed. Sovereignty is generally upheld at all costs (O'Reilly 2011).

[7] This was particularly noted several times in the United Kingdom, perhaps because of the long-cultivated British-Antarctic masculine mythos, underscored by the fact that the British Antarctic Survey did not allow women scientists or workers at their research stations until 1986, with no woman overwintering until 1997. The pace of change in terms of gender representation has been fairly rapid since the late 1990s. A woman now directs the British Antarctic Survey.

[8] The grounding line is the front wall of the glacier, usually where the foot of the glacier stops on land. WAIS has a marine grounding line; it is underwater. This changes how the grounding line behaves, but if a grounding line exhibits retreat, it may trigger disintegration of the entire ice sheet.

REFERENCES

BAMBER, J.L & W.P. ASPINALL 2013. An expert judgment assessment of future sea level rise from the ice sheets. *Nature Climate Change* **3**, 424-7.

BRYSSE, K., N. ORESKES, J. O'REILLY & M. OPPENHEIMER 2012. Climate change predictions: erring on the side of least drama? *Global Environmental Change* **23**, 327-37.

CAREY, M. 2010. *In the shadow of melting glaciers: climate change and Andean Society*. New York: Oxford University Press.

COLLINS, H. & R. EVANS 2007. *Rethinking expertise*. Berkeley: University of California Press.

CRUIKSHANK, J. 2005. *Do glaciers listen? Local knowledge, colonial encounters, and social imagination*. Vancouver: University of British Columbia Press.

DALKEY, N.C. 1967. *Delphi*. Santa Monica, Calif.: RAND Corporation.

EDWARDS, P.N. 2010. *A vast machine: computer models, climate data, and the politics of global warming*. Cambridge, Mass.: MIT Press.

GEURTS, K.L. 2002. On rocks, walks, and talks in West Africa: cultural categories and an anthropology of the senses. *Ethos* **30**, 178-98.

GRANDIN, T. & C. JOHNSON 2004. *Animals in translation: using the mysteries of autism to decode animal behavior*. Boston: Mariner Books.

GUSTERSON, H. 1996. *Nuclear rites: a weapons laboratory at the end of the Cold War*. Berkeley: University of California Press.

HANSEN, J.E. 2007. Scientific reticence and sea level rise. *Environmental Research Letters* **2**, 1-6.

HARAWAY, D. 1989. *Primate visions: gender, race, and nature in the world of modern science*. New York: Routledge.

————— 1997. *Modest_Witness@Second_Millennium. FemaleMan©_Meets_OncoMouse^{TM}: feminism and technoscience*. New York: Routledge.

HELMER, O. 1967. *Analysis of the future: the Delphi method*. Santa Monica, Calif.: RAND Corporation.

HELMREICH, S. 2009. Intimate sensing. In *Simulation and its discontents* (ed.) S. Turkle, 129-50. Cambridge, Mass.: MIT Press.

HERZFELD, M. 2005. Political optics and the occlusion of intimate knowledge. *American Anthropologist* **107**, 369-76.

HERZIG, R. 2005. *Suffering for science: reason and sacrifice in modern America*. New Brunswick, N.J.: Rutgers University Press.

HOWES, D. 2003. *Sensual relations: engaging the senses in social and cultural theory*. Ann Arbor: University of Michigan Press.

HSU, C. & B.A. SANDFORD 2007. The Delphi technique: making sense of consensus. *Practical Assessment, Research & Evaluation* **12: 10**, 1-8.

HUSTAK, C. & N. MYERS 2012. Involutionary momentum: affective ecologies and the science of plant/insect encounters. *differences* **23**, 74-188.

INGOLD, T. 2000. *The perception of the environment: essays on livelihood, dwelling and skill*. New York: Routledge.

IPCC 2013. *Climate change 2013: the physical science basis. Contribution of Working Group I to the Fifth Assessment Report of the Intergovernmental Panel on Climate Change* (eds T.F. Stocker, D. Qin, G.-K. Plattner, M. Tignor, S.K. Allen, J. Boschung, A. Nauels, Y. Xia, V. Bex & P.M. Midgley). Cambridge: University Press.

JOUGHIN, I., B.E. SMITH & B. MEDLEY 2014. Marine ice sheet collapse potentially under way for the Thwaites Glacier Basin, West Antarctica. *Science* **344**, 735-8.

KELLER, E.F. 1983. *A feeling for the organism: the life and work of Barbara McClintock*. San Francisco: W.H. Freeman and Company.

KNORR CETINA, K. 1999. *Epistemic cultures: how the sciences make knowledge*. Cambridge, Mass.: Harvard University Press.

KOHLER, R. 2002. *Labscapes and landscapes: exploring the lab-field border in biology*. Chicago: University Press.

KOHN, E. 2013. *How forests think: toward an anthropology beyond the human*. Berkeley: University of California Press.

LAHSEN, M. 2005. Seductive simulations? Uncertainty distribution around climate models. *Social Studies of Science* **35**, 895-922.

LATOUR, B. 1999. *Pandora's hope: essays on the reality of science studies*. Cambridge, Mass.: Harvard University Press.

————— 2013. *An inquiry into modes of existence: an anthropology of the moderns*. Cambridge, Mass.: Harvard University Press.

MYERS, N. 2009. Performing the protein fold. In *Simulation and its discontents* (ed.) S. Turkle, 171-202. Cambridge, Mass.: MIT Press.

NATIONAL RESEARCH COUNCIL 2007. *Exploration of Antarctic subglacial aquatic environments: environmental and scientific stewardship*. Washington, D.C.: The National Academies Press.

OPPENHEIMER, M., B.C. O'NEILL, M. WEBSTER & S. AGRAWALA 2007. Climate change: the limits of consensus. *Science* **317**, 1505-6.

O'REILLY, J. 2011. Tectonic history and Gondwanan geopolitics in the Larsemann Hills, Antarctica. *PoLAR: Political and Legal Anthropology Review* **34**, 214-32.

————— 2015. Glacial dramas: typos, projections, and peer review in the Intergovernmental Panel on Climate Change. In *Climate cultures: anthropological perspectives on climate change* (eds) J. Barnes & M. Dove, 107-26. New Haven: Yale University Press.

————— forthcoming. *Technocratic wilderness: an ethnography of scientific expertise and environmental governance in Antarctica*.

—————, N. ORESKES & M. OPPENHEIMER 2012. The rapid disintegration of projections: the West Antarctic Ice Sheet and the Intergovernmental Panel on Climate Change. *Social Studies of Science* **42**, 709-31.

ORESKES, N., K. SHRADER-FRECHETTE & K. BELITZ 1994. Verification, validation, and confirmation of numerical models in the earth sciences. *Science* **263**, 641-6.

ORLOVE, B. 2005. Human adaptation to climate change: a review of three historical cases and some general perspectives. *Environmental Science & Policy* **8**, 589-600.

PFEFFER, W., J. HARPER & S. O'NEEL 2008. Kinematic constraints on glacier contributions to 21st-century sea level rise. *Science* **321**, 1340-2.

RABINOW, P. 1996. *Making PCR: a story of biotechnology.* Chicago: University Press.

RAFFLES, H. 2002. Intimate knowledge. *International Social Science Journal* **54**, 325-35.

RAHMSTORF, S. 2007. A semi-empirical approach to projecting future sea-level rise. *Science* **315**, 368-70.

RIGNOT, E., J. MOUGINOT, M. MORLIGHEM, H. SEROUSSI & B. SCHEUCHL 2014. Widespread, rapid grounding line retreat of Pine Island, Thwaites, Smith, and Kohler glaciers, West Antarctica, from 1992 to 2011. *Geophysical Research Letters* **41**, 3502-9.

SHAPIN, S. & S. SCHAFFER 1989. *Leviathan and the air-pump: Hobbes, Boyle, and the experimental life.* Princeton: University Press.

SKRYDSTRUP, M. 2012. Modelling ice: a field diary of anticipation on the Greenland Ice Sheet. In *The social life of climate change models: anticipating nature* (eds) K. Hastrup & M. Skyrdstrup, 163-82. New York: Routledge.

STOLLER, P. 1989. *The taste of ethnographic things: the senses in anthropology.* Philadelphia: University of Pennsylvania Press.

TRAWEEK, S. 1988. *Beamtimes and lifetimes: the world of high energy physicists.* Cambridge, Mass.: Harvard University Press.

TSING, A. 2005. *Friction: an ethnography of global connection.* Princeton: University Press.

VAUGHAN, D.G. & J.R. SPOUGE 2002. Risk estimation of collapse of the West Antarctic Ice Sheet. *Climatic Change* **52**, 65-91.

Sentir la glace : science de terrain, modèles et intimité experte avec le savoir

Résumé

Cet article analyse les relations des glaciologues de l'Antarctique (scientifiques de terrain, modélisateurs et utilisateurs de données de télémétrie par satellite) ont nouées avec l'environnement glaciaire. Ces relations contribuent à comprendre la nature tactile et expérientielle de l'expertise scientifique, même lorsque les experts affirment ne pas détenir beaucoup de savoir scientifique. Leurs extrapolations à propos de la nature, en l'absence de données, suggèrent une intimité de ces scientifiques de terrain avec l'environnement dans lequel ils travaillent et des modélisateurs avec les modèles virtuels futurs avec lesquels ils interagissent. Mes recherches sur l'engagement sensoriel parmi les scientifiques révèlent également les manières dont ce qui est incarné et sensoriel n'est pas primitif, élémentaire, basique ou instinctif mais intégré dans les circonvolutions du nationalisme, des traductions scientifiques de l'échelle et des conflits aux marges de ce que chaque discipline scientifique considère comme son propre domaine d'expertise. L'expertise se crée dans les espaces entre les rencontres intimes, en relation avec le poids de l'apprentissage culturel qui enseigne aux aspirants experts comment rencontrer, analyser, comparer et interpréter.

3

Uncertainty in the signal: modelling Egypt's water futures

JESSICA BARNES *University of South Carolina*

The impact of climate change on the Nile holds great significance for Egypt, which sources almost all its water from the river. Yet the models used to forecast these impacts produce a range of results, spanning from projections of increased flow to the opposite. Such uncertainty, to varying degrees, is a central feature of debates over climate change impacts around the world, and can be put to a variety of political uses. While all scientists negotiate uncertainty in their everyday practice, I focus here on two particular groups: Egyptian and non-Egyptian scientists who are working on modelling Nile futures. First, I examine Egyptian scientists' underplaying of key data uncertainties and link this to their position within national expert networks as predictors of their nation's water futures. Second, I highlight a material dimension to scientists' ability to know, address, and reduce uncertainty. Finally, I argue that Egyptian scientists' selective interpretation of uncertain model outputs, focusing on results that indicate a declining flow, is part of a political effort to bolster Egypt's claim on Nile waters. By exploring uncertainty as a known entity that scientists – not just in Egypt, but in any political context – actively manage, the paper highlights the linkages between the uncertainty of knowledge about various futures and the material, cultural, and political positions of those who produce that knowledge.

It is a November day in 2011 and several hundred policy-makers, water managers, and scientists from the Middle East and beyond have gathered in Cairo for the Second Arab Water Forum. In the conference room of a five-star hotel, a scientist from Egypt's Ministry of Water Resources and Irrigation stands behind the podium. She presses a key on the computer and brings up her next PowerPoint slide. On one side of the slide there is a map of the Nile Basin, with a bold arrow pointing to a site along the curvy blue line of the Nile River in northern Sudan. This is the town of Dongola, one of the main hydrological monitoring stations along the Nile. Next to the map, there is a graph. The x-axis shows the months of the year. The y-axis is labelled 'flow changes' and ranges from −50 per cent to +50 per cent. The scientist explains that the multiple wiggly lines on the graph indicate the results of her team's computer modelling studies, which project Nile flows at Dongola in the year 2050. Their conclusion is that the annual flow of the Nile forty years from now will be between −13 per cent and +36 per cent of its historical average, owing to climate change.

Journal of the Royal Anthropological Institute (N.S.), 46-66
© Royal Anthropological Institute 2016

As the source of over 95 per cent of Egypt's water, the Nile, and its future, holds fundamental significance for Egypt. In the scientist's presentation at the water forum, she performs a particular type of knowledge about the Nile – one rooted in computer modelling. A central feature of this knowledge is its explicit uncertainty. As the scientist acknowledges, the ministry's models have produced a wide range of results. Increasing concentrations of greenhouse gases in the atmosphere are changing precipitation patterns in the Nile's East African source regions, but scientists do not know exactly how these patterns are changing and so whether river flows are likely to increase or decrease in coming decades.[1]

In this paper, I examine the making and imagining of a Nile future shaped by climate change. In doing so, I shift my focus away from the farmers and irrigation engineers who interact very physically with this water on an everyday basis (see Barnes 2014), and towards the Egyptian and non-Egyptian scientists who interact with this water, instead, through the equations of computer models, figures of billions of cubic metres, graphs published in academic journals, and the dense language of statistics.[2] The modelling work that these scientists do provides insights into how Egypt's water supply might change if the globe warms. Such information has important implications for how water is managed today, both within Egypt and within the broader Nile Basin.[3] The way in which various scientists approach these models and the uncertainty associated with them therefore matters since it shapes the message they communicate to broader publics. This paper draws on interviews I conducted with Egyptian and non-Egyptian scientists[4] between 2012 and 2014 and participant observation at several international water meetings.[5]

My central interest is in the production and management of uncertainty surrounding forecasts of resource futures. Uncertainty, and the way in which those who generate and interpret scientific knowledge may put it to political use, is a widespread and essentially irreducible feature of debates over climate change impacts around the world. By focusing on two groups of scientists and their work to model the impacts of climate change on a critical resource, I seek to identify how material and political relations influence the response of scientists in different places – not just in Egypt, but in any political context – to uncertainty in data, modelling processes, and model outputs. I argue that the uncertainty in the signal – namely, whether or not climate change will lead to an increase or decrease in Nile flows – is shaped not only by theories of climate change and correlations between precipitation and discharge. It is shaped also by technologies and sociopolitical relationships that influence scientists' abilities to produce models of climate-river dynamics in the first place, and by scientists' willingness to present particular model outputs to particular audiences.

Uncertainty, and the ways in which it may be generated, assessed, negotiated, and acted on by situated actors, has long been a theme within anthropology (e.g. Appadurai 1998; Boholm 2003; Douglas & Wildavsky 1983; Johnson-Hanks 2005; Murphy 2006; Rapp 2000; Taddei 2012; Wright & Phillips 1979). Anthropologists have shown how uncertainty about public health threats, for instance, may legitimate government interventions in the name of preparedness (Lakoff 2008), be tied to the emergence of various coping technologies (Samimian-Darash 2013), play into scientifically inspired prophecies (Caduff 2014), or constitute a foundation for citizens' claims-making (Petryna 2002). In these ways, uncertainty can serve particular political and discursive ends. Yet it may also undermine the legitimacy of some actions, as in the case of

Journal of the Royal Anthropological Institute (N.S.), 46-66
© Royal Anthropological Institute 2016

humanitarian work, for example, which is predicated on the certitude of an emergency and a moral imperative to act (Redfield 2013).

Building on the idea that uncertainty may open up or foreclose certain political possibilities, and thus that its generation and representation are highly significant, I focus here on a particular kind of uncertainty: scientific uncertainty. By this I mean not uncertainty in a colloquial sense, as in something that is not known for sure, but uncertainty as a technical term whose contours (in the minds of scientists) may actually be quite certain. To scientists, uncertainty refers to imprecision in the outcome of a modelling exercise or experiment; it can be measured and classified using tools of statistical probability (O'Reilly, Brysse, Oppenheimer & Oreskes 2011). This close definition of what uncertainty means to a scientist in the context of his or her work does not mean that this is the only form of uncertainty that a scientist engages with. Indeed, scientists often move from talking about statistical analyses of uncertainty to casual usage of the term in relation to things they are not sure about. Furthermore, a number of other uncertainties, from uncertainty about where the next grant will come from to uncertainty about the security of a research position, may affect the choices that scientists make when running their models and interpreting the results. By directing ethnographic attention to scientists' ways of knowing uncertainty as a central constituent of their scientific work, however, this paper elucidates some key moments in the production of uncertainty and links these moments to the material, cultural, and political position of those who make that knowledge.

The paper speaks, therefore, to broader concerns within anthropology about the production and interpretation of scientific knowledge. A number of anthropologists have looked at how scientific knowledge is generated in both laboratory and field settings (e.g. Gusterson 1998; Helmreich 2009; Knorr Cetina 1999; Latour & Woolgar 1986; Traweek 1992). This work has clearly demonstrated how scientists negotiate complex positionalities in the process of producing knowledge. It has also brought to the fore uncertainty as something that all scientists constantly face and manage in a variety of ways (Star 1985). Through an examination of how climate change projections are made and used in a national science context, this paper offers insights into how scientists negotiate political terrains and disciplinary expectations as they talk about their data and results, and how these terrains are also bounded by material constraints.

First, I look at Egyptian scientists' response to the uncertainty regarding a central piece of data: the general circulation models' predictions of future precipitation over East Africa. I show how this response is influenced by the position these scientists hold within national policy and expert networks as predictors of their nation's water futures.[6] Second, I draw attention to the materials that shape the production of climate change knowledge. I analyse how these materials impact scientists' ability to know, address, and reduce uncertainty. Third, I explore how the political context influences how scientists navigate and utilize the uncertainty surrounding model outputs. I demonstrate how Egyptian scientists tend to focus selectively on an outlook of reduced Nile flows, as part of an effort to maintain Egypt's claim on Nile waters. To contextualize this analysis, in the next section I situate climate change alongside another important factor that will affect the future of Egypt's water supply: withdrawals by upstream states along the Nile. I also introduce the Egyptian and non-Egyptian scientists who are working on projecting climate change impacts and the series of models that they employ to do this predictive work.

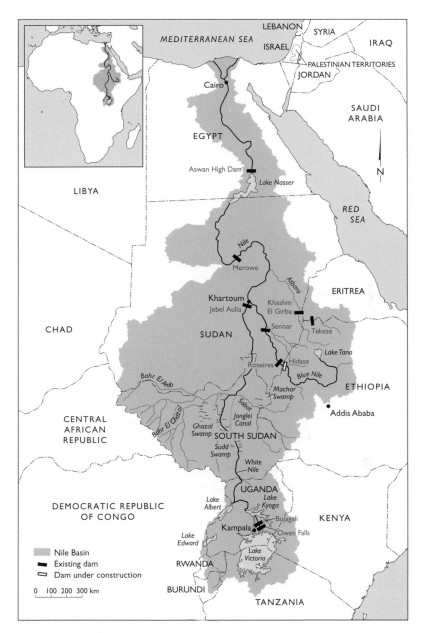

Figure 1. The Nile Basin (Credit: Bill Nelson).

Nile futures

Egypt has long been a dominant player within the Nile Basin, despite its position at the downstream end of the river (Fig. 1). As a country whose name is almost synonymous with the Nile, Egypt has vigorously defended its rights to the waters of this river, based on its historical usage of the Nile flood for irrigation. These rights are enshrined in the 1959 Agreement for the Full Utilization of Nile Waters, which allocates the water of the river between Egypt and Sudan alone, Egypt receiving a

Journal of the Royal Anthropological Institute (N.S.), 46-66
© Royal Anthropological Institute 2016

much larger share. Recent decades, however, have seen increasing challenges from the other ten states that share the river basin and wish to utilize its waters (Cascão 2008; 2009; Collins 1994; Waterbury 1979; 2002). A new legal agreement, the Cooperative Framework Agreement, has been signed by six states and, as of January 2016, ratified by three of those states. If the agreement comes into force, it will open up the possibility for upstream countries to increase their withdrawals from the river (Mekonnen 2010; Nicol & Cascão 2011). In a move that signifies its determination to challenge the status quo, Ethiopia also began work in 2011 on a major new dam: the Grand Renaissance (Hidase) Dam. Situated across the Blue Nile, which contributes 85 per cent of the main Nile's discharge, the dam has the potential to profoundly alter Sudan and Egypt's water supply. This is the geopolitical context, therefore, in which scientists have begun to explore the significance of climate change for shaping the future of this shared resource.

Model experts

A number of scientists around the world who work on climate change have taken the Nile Basin as a focal point for their research (e.g. Bhattacharjee & Zaitchik 2015; Conway 2005; Conway & Hulme 1993; 1996; Gleick 1991; Kingston & Taylor 2010; Strzepek, Yates & El-Quosy 1996; Yates & Strzepek 1998). In some cases, these scientists, whom I refer to as 'non-Egyptian scientists', began their work on the Nile as consultants on one of the donor-funded projects that have examined climate change impacts in the region. In other cases, it was their own research agenda that led them to the Nile. A heterogeneous group, some of these scientists have a degree of attachment to the region having spent time in various Nile Basin countries and worked closely with local people. Other scientists have never visited; for them, the Nile Basin is purely an object of study. While the geopolitical context of the shared Nile Basin may be a source of interest, to most of these scientists from outside of the region, it is not something in which they have a direct stake.

The geopolitics of the Nile Basin holds a different meaning, though, to the group of around ten Egyptian scientists with Ph.D.s who are also working in this field. Many of these Egyptian scientists are employed, or have in the past been employed, by Egypt's Ministry of Water Resources and Irrigation (hereafter 'the ministry'). This ministry is responsible for managing the network of canals, dams, and pumping stations that control the distribution of Nile water through the country. The ministry has several modelling-focused units and an associated research centre, whose role is to generate data about current and future river flow patterns to inform both contemporary management decisions and strategic planning. Other Egyptian scientists work for regional organizations, like the Nile Basin Initiative, or are academics at Cairo University. Some started working with these models for their graduate studies; others began when they were working for donor-funded projects and learned how to use the models through training workshops. In many cases, their continued engagement with these models is part of their job as employees of the water ministry, which has dedicated a small staff to exploring climate change impacts. The task of these researchers is to produce knowledge about the future that will, in theory, inform water distribution decision-making and, potentially, climate change adaptation activities. For example, future flow projections could influence the ministry's policy on how much water to store each year in its main reservoir, Lake Nasser, and how much to release through the gates of the Aswan High Dam; the area it permits to be cultivated with water-intensive crops like rice; or the degree to which it promotes the development of alternative water sources, such as deep groundwater or desalination. While it is not evident up till now that these data have had a transformative effect on day-to-day policy-making within the water ministry, there is certainly the potential for this knowledge about the future of the Nile to shape the way in which ministry officials think about water distribution.

Two brief biographies are illustrative of the pathways that have led Egyptian scientists to this work. Dr Nizar is one of the most internationally prominent of the Egyptian scientists, having published his work both in English-language peer-review journals, like the *Journal of Hydrologic Engineering*, as well as in regional outlets such as the *Nile Basin Water Engineering Scientific Magazine*. He started his career as a hydrologist in Egypt's water ministry. When I interviewed him over the phone in January 2014, he explained how his interest in climate change and the Nile dated back to the late 1990s. 'I started reading about this and found very little done on the Nile, at least by Nile people, Nile inhabitants', he told me. This spurred him to a research topic for his Ph.D., which he did in the United Kingdom, and a postdoctoral position in the Netherlands. For a number of years he worked in the ministry's Nile Forecast Centre. This is one of the main centres of climate change work in Egypt. Nizar has since moved overseas and is working in an office of the Nile Basin Initiative, modelling the impact of climate change on different infrastructure within the basin.

Like Nizar, Dr Mona, the scientist whose presentation I described at the start of this paper, also began her career in the ministry. She conducted her Ph.D. research at Ain Shams University in Cairo on techniques of rainfall estimation over the Nile Basin. While not formally trained in climate modelling, she has worked for many years at the Nile Forecast Centre. Her name has yet to appear on a scientific publication about climate change, but having been involved in a number of donor-funded climate change projects, she is quite knowledgeable about the topic, hence her selection to present the results of the research group to the Arab forum. She tells me that she plans to focus on climate change in the future 'to get some research and complete my career'.

Cascades of models

To probe these scientists' varied relations with model inputs, processes, and outputs, we must first understand the models that they use to translate greenhouse gas emissions into changing river flows. The initial step in what scientists refer to as a 'model cascade' is to get data from general circulation models (GCMs). These models comprise a series of equations written in computer code, which represent the physical dynamics of atmospheric-land-ocean interactions over a globe divided into three-dimensional grid cells. 'Running' a GCM involves inputting the starting conditions into a supercomputer, choosing various parameters, and the computer then processing these equations for each grid cell for iterations of time into the future. This can only be done at the twenty-three research centres around the world that have the technical and computing capacity to develop and operate a GCM.[7] The GCMs produce a range of outputs on key climatic variables for the global grid, which the research centres then make available (to a greater or lesser degree) for other scientists to use.

One grid cell in a GCM can, however, cover anything from 250 km to 600 km. In the case of the Nile Basin, for instance, this would mean that in a coarse-resolution GCM, Khartoum, which lies at an elevation of 386 m and has an average annual precipitation of 155 mm, could fall within the same grid cell as the Ethiopian Highlands, where elevation rises up to 4,500 m and annual rainfall can be ten times that amount. An average figure for this grid cell would erase this contrast. Thus the next step in the modelling process, which is called downscaling, is to obtain data that are more detailed over space.[8] This is most commonly done through the use of regional climate models (RCMs).[9] With a regional model, scientists are able to bridge the gap between the GCM output (typically 250 km) and the kind of data that are needed to drive a hydrological model (typically 20 km). A number of modelling institutes around the world have developed regional climate models, including the UK Met Office Hadley Centre and the International Centre for Theoretical Physics in Italy.[10] Some of these models are freely available and can be modified by users; others are commercial.

Finally, to translate regional data on future climatic variables into a figure for Nile water futures, scientists employ hydrological models. These models break down river basins into a series of grid cells and model interactions between them according to a set of physical equations and parameters, which represent the flow of water over and through the soil. Egyptian scientists use the Nile Forecast System – a model developed by international consultants through US-funded projects in the 1990s. Scientists working outside of Egypt use other kinds of hydrological models as the Nile Forecast System is not publicly available.

Through use of this model cascade, scientists are able to envisage Nile futures. As one scientist explained to me, 'Most of the river upstream [of Egypt's Aswan High Dam] is not managed, so starting from the climate conditions you can have a reasonable guess [about how much river flows will be]'. By 'not managed' he means that the flows are not completely controlled by dams, and so there is a direct relationship between climate and discharge. Downstream of the Aswan High Dam, on the other hand, the water is 'fully managed'. This means that the flow of water through Egypt is tightly controlled by a series of dams and weirs (Barnes 2014). Thus we are talking here about a *national* prediction of how much water Egypt will receive in the future, rather than a prediction of how that future will be experienced by people living in Egypt (Barnes 2015). This is a prediction, however, that is scientifically uncertain. Despite the scientific care put into taking GCM data, running a RCM, and inputting the results in a hydrological model, the visioning of Nile futures remains ultimately, as the scientist quoted above acknowledged, a matter of 'reasonable' guesswork.

Uncertainty in the signal

At several academic conferences I have attended in the United States, I have talked about this research project. Like the Egyptian scientist at the Arab forum, I have stood in front of an audience and, using the sentence I used above, explained that modelling studies suggest 'that the annual flow of the Nile forty years from now will be between −13 per cent and +36 per cent of its historical average'. In each case, the audience has responded with laughter, although my intent was not to make a joke. To many social scientists, perhaps this kind of result seems slightly ridiculous. A prediction that Nile flows may increase or may decrease seems meaningless. But to climate scientists, it is not ridiculous, nor is it all that surprising. They understand that uncertainty accompanies all climate projections. They know that precipitation-related projections are among the most uncertain, and that doubt over whether future rainfall will be higher or lower is not uncommon. They are probably at least somewhat familiar with the voluminous scientific literature on this precise topic (e.g. Curry & Webster 2011; Mearns 2010). Indeed, in the case of the Nile, although the figures and ranges vary, a consistent theme

Journal of the Royal Anthropological Institute (N.S.), 46-66
© Royal Anthropological Institute 2016

among all the studies of climate change impacts on the river is that the results span from the positive to the negative (Beyene, Lettenmaier & Kabat 2010; Conway & Hulme 1996; Elshamy, Seierstad & Sorteberg 2009; Kingston & Taylor 2010; Nawaz, Bellerby, Sayed & Elshamy 2010; Soliman, Sayed & Jeuland 2009; Strzepek *et al.* 1996; Taye, Ntegeka, Ogiramoi & Willems 2011). Scientists refer to the possibility of an increasing or decreasing trend as 'uncertainty in the signal'.

Garbage is in the eye of the beholder

Over lunch one day I talked with a scientist who works on climate change impacts and adaptation in a region of the United States where, as in East Africa, there is similar lack of agreement among the GCMs on whether precipitation is likely to increase or decrease. 'I've been thinking of a call centre for downscaling which people could call in to', she told me. 'The voice at the end of the line would ask, "What do you want, precipitation or temperature?" If the person wants precipitation data, they would be told to press a button. Then, the recorded message would say, "I'm sorry, that data is not available, please call back in forty years"'. We laughed. 'But don't people need this kind of data, don't they need precipitation data on a small scale?' I asked her. 'Well I'm pretty sure that if people knew what it is [i.e. its unreliable quality] they wouldn't want it', she responded. To this American academic, the downscaled precipitation data for her region are sufficiently dubious that in her opinion they are basically useless.

The fact that GCMs have a limited ability to reproduce precipitation, owing to its high spatial and temporal variability and non-linear nature, is widely acknowledged in the literature (e.g. Bloschl & Montanari 2010; Hawkins & Sutton 2011; Kundzewicz & Stakhiv 2010). If the GCM precipitation data are unreliable, then the data that are fed into the downscaling model are also unreliable and the whole modelling enterprise is doomed. None the less, contrary to this academic's opinion, even if some people do understand the complex, flawed nature of these data, they still want them. To understand why this is the case, we must look at the political and policy contexts in which scientists are working – contexts that may have different time horizons to those of this scientific research. Since no data are ever absolutely certain, at root, here, are divergent assessments about what makes data 'good enough' for a particular purpose.

In March 2012 I interviewed a scientist in the United States, who specializes in climatic and hydrologic variability, about his work on the Nile. 'For Ethiopia [where 85 per cent of the Nile flow is generated]', he said, 'you can't believe any of these [general circulation] models. If you can't believe what they say today, you can't believe what they say tomorrow. Taking something that is wrong at a large scale and then looking at a fine scale, I don't know . . . ' His voice petered out, the implication being that such work was not all that useful. 'I've come to the conclusion that a lot of the literature [that uses these data] is just wrong'. He explained how there are systematic problems in how the GCMs depict fluxes of water vapour off the Indian Ocean, which drive precipitation in the East Africa region. If the data that you put into the domain of your regional model are so faulty, he concluded, 'then you're kinda cooked'.

This perspective that the GCM precipitation data are too uncertain to be of use was something I heard from a number of the non-Egyptian scientists to whom I talked. One described how he had observed many hydrologists over the past ten years taking GCM output and running hydrological models, but that he saw it as a waste of time. 'With rainfall [data from GCMs]', he said, 'I wouldn't call it uncertain; it's just inaccurate. Sometimes, we say garbage in, garbage out'. Without a sound comprehension of circulatory processes, it is impossible to make good equations ('garbage in'), and so the quality of the precipitation data produced is questionable ('garbage out').

However, when I asked Egyptian scientists about whether they consider the GCM data for East African rainfall to be problematic for downscaling work, I heard a different response. While they acknowledged that GCMs can have problems modelling precipitation, they expressed confidence that the GCMs they had selected to work with were reliable. 'Actually the data we use from ECHAM and Hadley [two GCMs] is good, so maybe this problem is for other areas', one told me. Another commented, 'Definitely there exists some different results from all these models'. But in terms of the model they are using, 'It works, *really*', he said. Such perspectives are important because they validate the continuing work to downscale these GCM results, use them to drive hydrological models, and produce knowledge about the future of the Nile.

Notably, these scientists were not making this contrasting judgement of the GCM precipitation data on the basis of a scientific evaluation – a process that scientists refer to as 'evaluating model skill'. Rather than conducting such an evaluation – a complex process to which whole research groups are dedicated – most scientists who work on downscaling choose data from GCMs that they have heard work well for their regions or parameters of interest. This selection is guided either by peer-review publications or by more informal communication between scientists – a conversation at a meeting, for example, in which someone says, 'Oh, that model runs cold for this area', or 'The Hadley model works well for that area', as another scientist explained to me.

So, given that the Egyptian scientists, like most scientists who work on downscaling, have not done in-depth studies to evaluate the GCMs they are working with, what explains the higher level of confidence they express in the East African precipitation projections? It is not uncommon for scientists to make accommodations about what data they deem acceptable as a result of their broader research agendas, and it is certainly not a practice limited to Egyptian scientists alone. Other studies have clearly shown, for example, how climate scientists portray the uncertainty of their modelling work in different ways depending on their audience (O'Reilly *et al.* 2011), in some cases underplaying uncertainty (Lahsen 2005), in other cases foregrounding it (Brysse, Oreskes, O'Reilly & Oppenheimer 2013; Shackley & Wynne 1996).[11] In this instance, I argue, the confidence that Egyptian scientists express in the GCM precipitation data stems from the position they hold as predictors of their nation's water futures.

Many of the Egyptian scientists work for the Ministry of Water. Given that their job is to predict the future of the Nile, they have only limited freedom to choose what they want to research. This is a point of contrast with non-Egyptian scientists, a number of whom told me that the uncertainty of the GCM precipitation data had led them to work on something else. The American scientist who warned that we are 'kinda cooked' is no longer working on modelling climate change impacts in the Nile Basin. A British scientist who formerly worked on the Nile explained that although he started out his research career working with the output from GCMs to do climate change impact assessments, he had chosen a new research direction. He recounted, 'I decided that because of the uncertainty, I'd move away to adaptation, which is what I primarily focus on nowadays'.[12]

For Egyptian scientists who work for national research institutes, on the other hand, their career progression is based on them continuing to do the modelling work that their superiors within the ministry deem valuable. In these circumstances, they have an interest in presenting the GCM precipitation data as adequate. This is not to say that

Egyptian scientists are somehow less rigorous than other scientists, or more willing to accept poor data for purely instrumental reasons. Rather, it is illustrative of a wider dynamic of how scientists may put uncertainty to productive use in a variety of ways. Arguing for the quality of the GCM data that feed the scientists' regional hydrological models is a way of arguing for their own predictive skill. Without 'knowing' how much rain is going to fall in the East African Highlands, it is impossible to 'know' how much water is going to flow down the Nile. This perspective was evident in the response that one Egyptian scientist gave when I asked whether it made sense to work with such problematic data: 'But you have to do something, even if it's not very reliable. You have to do something'.

The motivation driving Egyptian scientists to continue working in this field is not, though, only one of obligation linked to their employment. A number of the scientists whom I interviewed expressed their commitment to working on an issue that potentially holds great import to Egypt.[13] When I asked these scientists what led them to research in climate change, many narrated a story that linked their work to a national need. One told me, for example, 'As you know, Egypt suffers from water scarcity and we have different problems regarding water shortage. We need to quantify the impact of climate change on the water supply for Egypt. So I tried to help my ministry and my country'. Another scientist said that he did his Ph.D. work on another topic but switched to climate change modelling, 'because it is now much more important in our region. I am interested in the community in Egypt. Egypt will be suffering much from the impacts of climate change'.

At play here is both a sense of opportunism – seeing a gap in the research that can be filled – and a sense that it is important to have more knowledge about an issue of national significance. One Egyptian scientist whom I interviewed told me: 'When I did my Ph.D., no one had touched these areas … I was interested actually to look at the impacts on Egypt because we have the only source of water in the Nile and I wanted to see how our resource would be affected and how much we are at risk'. So there is a desire for data on resource futures that are produced closer to home. These scientists imply that data on Egypt's water futures produced by Egyptians are fundamentally more valid.[14] Whether or not other people actually see this science as being more trustworthy is another matter, for as Lahsen (2007) points out, a range of sociocultural and political dynamics shape how policy-makers and scientists perceive the findings of their fellow nationals. The key point, however, is not so much whether this claim is true, but rather that it provides a motivation for Egyptian scientists to continue working with GCM data – to take those data as 'good enough' – so as to produce Nile forecasts.

In this context, therefore, it is not surprising that Egyptian scientists should present a lesser degree of uncertainty about the East African precipitation projections from the GCMs that they use, in contrast to some of their foreign counterparts' sweeping dismissals. While they know the limitations of these data, they have a strong interest in continuing to use them and therefore, at least publicly, in defending their validity.

Disks, processing power, and power cuts
Yet Egyptian scientists know that the output of their predictive models will always be uncertain. One scientist told me about the first climate change modelling project that the ministry was involved in, which was funded by the Dutch. 'They had a very wide output [for Nile flow]', he reported. 'It was very large – +30 to −25 [per cent]'. The fact that he could remember these figures from a project that ended nine years

previously is indicative of their fundamental significance. He continued, explaining that this uncertainty is one reason for the ministry's ongoing investments in modelling: 'After that the ministry tried to know more exact results about the climate change. Whether there will be change or no. Plus or minus. They need more narrow data'. Thus one of the motivations for further research is a desire to reduce this uncertainty, the implication being that a more precise forecast will facilitate policy-making. Although there is ample evidence that more numbers and lower uncertainty do not necessarily lead to better decision-making (e.g. Sarewitz, Pielke & Byerly 2000), and that ignorance, or a lack of certainty, may in itself be 'usable' (Ravetz 1987), the assumption that narrowing the uncertainty bandwidth is a valuable goal is widespread among scientists working on Nile flow predictions. The degree to which this uncertainty is reducible, though, depends on where that uncertainty is coming from.

Scientists identify three main sources of uncertainty when it comes to projecting the impacts of climate change (Hawkins & Sutton 2009; 2011). First, there is the natural internal variability of the climate system. In other words, there are natural fluctuations that are unrelated to radiative forcing, which cannot be captured by climate models. Second, there is uncertainty about the trajectories of future emissions of greenhouse gases and aerosols. The Intergovernmental Panel on Climate Change (IPCC) has produced a set of scenarios, but nobody knows with certainty the quantity of greenhouse gases that the countries of the world will emit.[15] Third, there is uncertainty linked to the models. If you take two models and give them the same exact starting conditions and emissions scenario, they will produce different results. This is because of the contrasting ways in which different GCMs address processes that occur on a sub-grid scale and other kinds of adjustments that modellers make.

The first source of uncertainty is irreducible. There will always be internal variability in the climate system that no mathematical model can capture. But the second and third sources of uncertainty can be addressed through looking at multiple emissions scenarios, multiple models, or multiple runs of the same model (with different initial conditions or parameters), and comparing the results. While this does not reduce the uncertainty *per se*, it offers insight into the range of uncertainty and the likelihood of different outcomes. In some cases, scientists average the results. In the graph on Dr Mona's PowerPoint slide at the Arab Water Forum, for example, the green wiggly lines were interspersed with a single bold blue line. This line indicated the average of all the model simulations, and, in this case, lay above the zero point, giving an average picture of increased Nile flows in the future. On other occasions, scientists use complex statistical methods to compare the results.

This process of reducing uncertainty through the use of multiple simulations is, however, contingent on certain material needs. As Paul Edwards cautions in his book on climate modelling, 'When one is caught up in the cascade of words, images, and numbers, in the frenetic traffic from screen to screen, it is easy to lose sight of the infrastructure – to forget that, underneath that glistening surface of free-flowing information, computing remains a material process' (2010: 83). Hence, just as a substantial body of ethnographic work in laboratory settings has highlighted the role of material-technical artefacts in the production of scientific knowledge (e.g. Knorr Cetina 1999; Latour 1988; Traweek 1992), so, too, certain materials are enrolled in the production and reduction of uncertainty.

The first requirement for conducting multiple simulations is access to data. The GCM data are largely available for free in the public domain; the problem is how

Journal of the Royal Anthropological Institute (N.S.), 46-66
© Royal Anthropological Institute 2016

to get hold of them.[16] For Egyptian scientists to access these data, they must first be uploaded to the Internet from the modelling groups' hard drives, then downloaded over the Internet and saved on the Egyptian scientists' computers. For each GCM run, emissions scenario, and time period, globally gridded files for a range of parameters are required for input into the RCM. Every time you add another GCM, a different run of the same GCM, or another emissions scenario for that same time period, the number of files required increases accordingly. The combined size is huge. We are not talking here about the kilobytes and megabytes of text and image files, or the gigabytes of external hard drives, which readers of this journal might be most familiar with. We are talking about terabytes and petabytes of data. The size of the data means that they can take an extraordinarily long time to download – days or even weeks and months. This is especially the case in a country like Egypt where the mean download speed is 5 per cent that which it is in the United States, and where scientists' computers typically have relatively limited processor speed and memory.[17] 'Moving a terabit of data around is not very easy', one scientist working in the United Kingdom explained to me. 'We are used to doing it here in the UK, but in Egypt, they aren't'.

One scientist in the ministry identified this form of what Edwards terms 'data friction' – 'the costs in time, energy, and attention required simply to collect, check, store, move receive, and access data' (2010: 84) – as a key constraint. 'You know there is big difficulty to use more than two or three GCMs', she told me, 'because of the computational resources, also the data availability. It is freely available but it takes time to download'. She explained that this is a sufficient constraint that they often use an alternative strategy. 'Sometimes we depend on some person going to ICTP [the International Centre for Theoretical Physics in Italy] and bringing data on a hard disk because it takes too much time to be downloaded'.

When it comes to the regional climate modelling step, rather than needing to access data that have been produced elsewhere, Egyptian scientists are able to run the models themselves on their high-performance computers (purchased with funds from donor projects). However, there are still the material limits of what Edwards terms 'computational friction' – 'not only the physical and economic limits on processor speed and memory capacity, but also the human work involved in programming, operating, debugging, and repairing computers' (2010: 84). Running a model is not just about having computers with sufficient processing power and memory. It is also about the smooth operation of those computers not being interrupted by bugs in the computer code or breaks in the power supply. In addition, once the model produces its output, considerable time and expertise has to go into writing the programming sentences needed to process the model output into a useable format. One Egyptian scientist recounted her experience to me: 'During preparation of my Ph.D. I had to leave the computer running continuously for three months – three weeks for each scenario. If there was any glitch in the middle you would have to restart... Sometimes the model crashes or there is a cut in the electricity'.

As a result of such material constraints, which are faced to varying degrees in different research contexts, scientists have to compromise. For example, they might only use a single emissions scenario, or data from a single GCM run. They are well aware that they would get better purchase on uncertainty if they were to look at multiple combinations and scenarios, but material factors limit them. The scientist who told me about her Ph.D. experience, for instance, explained that she only looked at one scenario. 'It was

Journal of the Royal Anthropological Institute (N.S.), 46-66
© Royal Anthropological Institute 2016

too hard to run more scenarios because the computational resources were limited', she said.

So there is a material dimension to scientists' ability to conduct multiple simulations in order to gain a deeper understanding of the uncertainty surrounding a climate change prediction. It involves a set of human and nonhuman devices: the modeller and the computer programmer; an Internet connection; the electric charge that powers a computer; and hard disks travelling in the luggage of a scientist flying internationally. All of these factors shape the picture of the Nile's future and the degree of faith that scientists can attach to that picture.

Dealing with an uncertainty bandwidth

After Dr Mona's presentation to the Arab Water Forum there was a question-and-answer period. A man in the audience raised his hand. 'Do you think the water of the Nile will increase or decrease?' He paused, then added, 'Of course this is a big question, but what do *you* think?' Mona did not reply immediately, as other questions were being received from the audience. But when the panel had a chance to respond, another scientist from the ministry, also on the panel, spoke on her behalf. He refused to budge on the statement of uncertainty. 'In terms of the Nile, we don't know if flows will increase or decrease', he said.

The uncertainty over the future of the Nile is embodied in the range of the predictions that come out of the models. Scientists refer to this as the 'bandwidth'. It is the distance between the highest prediction (the highest possible future flow) and the lowest (the lowest possible future flow). Yet as the man in the audience's question illustrates, many water managers would like a definitive prediction. 'The policy-makers don't know how to deal with ranges', one Egyptian scientist commented to me at an international water conference. 'They want a figure'. In contrast, he said, 'We scientists in the ministry know uncertainty'. They know that it is impossible to give a definite answer. As another Egyptian scientist remarked to me, clearly thinking that I was implying that this uncertainty was a reflection of their limited modelling capacity, 'No one model in the world can give you one concrete figure of what will happen . . . even in the States'.

This raises the question of how different people respond to this bandwidth. In the face of a range, which part of that range do they focus on? When there is a spread, do they talk about the range, or, given the need for a policy response, do they focus on responding to one part of that range? What do officials do with an uncomfortable truth that climate change may be good for their country? Other work has demonstrated how politicians and policy-makers often choose among future scenarios for instrumental purposes, to promote particular policy agendas or build cross-party coalitions (van der Steen 2008). So, in the Egyptian case, how do scientists portray a threat that the Nile may increase or decrease in flow? Do scientists evenly portray flooding versus drought risk or do they emphasize one over the other?

Scientists assess uncertainty bandwidths in terms of probabilities. If more models suggest a particular signal – that is, a change in one direction – then that outcome is more likely. As one scientist explained to me, 'Nowadays you tend to use a lot more model outputs so you can attach some confidence. If ten are saying it is going to get drier and one says that it will get wetter, it will probably get drier'.[18] In other words, by weighing all the models, scientists are able to attach a language of confidence to their findings – a language of whether a particular outcome is 'very likely', 'likely', or 'not likely' that non-scientists can comprehend in disassociation from the numerical probabilities on which it is based.

So what do most models say about the future of the Nile? When I spoke to one non-Egyptian scientist about the results of his research group's modelling studies through 2050, he told me, 'While there was uncertainty, it was looking quite good for Egyptians. Most models showed a general increase in the flow of the river – broadly consistent with what other models say. We do expect an increase in the next few decades'.[19] A paper by Nawaz *et al.* made a similar statement: 'The climate change scenarios derived in this study were broadly in line with other studies, with the majority of scenarios indicating wetter conditions in the future. Translating the impacts into runoff in the basin showed increased future mean flows'. The authors concluded that 'managers therefore need to continue to prepare for the possibility of more frequent floods' (2010: 137).

But most of the Egyptian scientists I spoke to did not give me the same positive impression of increased flows for a water-scarce Egypt. Some, like the panellist at the Arab forum (whose stance may have been affected by the public setting), refused to give a specific answer. More commonly, though, they painted a picture of a declining water supply. As the scientist who heads one of the modelling units in the ministry told me, 'We are working on [the assumption that] there might be a decrease in the water arriving in the High Aswan Dam by about 20 to 25 per cent, depending on what kind of emission scenario'. So they are working on a presumption of a decrease, even though that same unit's modelling studies suggest that flows could increase under climate change.

In a number of instances, scientists explained this outlook by referencing the other kinds of futures that will interact with the climate futures they model – namely, futures of population growth and upstream states' plans to access more water. As one told me, 'The optimistic scenario shows there will be an increase in Nile flow of about 14 per cent. This will not solve the issue very much because demand will increase'. In a similar vein, another commented, '[With] climate change in thirty, forty, fifty years, even if we get increase [in river discharge], we have to look at development upstream in the basin because it may increase'. While the assumption that growing populations and upstream developments will raise demand obscures important questions about what the water is being used for (agriculture consumes much more water than people or hydroelectric dams do) and how it is being used (efficiency improvements will affect the degree to which demand increases), it remains widespread. By introducing such non-climatic variables, Egyptian scientists are able to maintain uncertainty in the face of a potential likelihood of a non-crisis (increased flow) future.

But why would they want to do so? This could, in part, be a pragmatic decision to look at and plan for a scenario that would exacerbate contemporary problems of water scarcity. In my opinion, though, the reason lies more in the geopolitical context and political sensitivity of information on Nile flows. Egypt is locked in a conflict with its upstream neighbours. It is concerned about how both Ethiopia's new dam and the new Nile treaty could adversely affect its current use of and right to Nile waters.[20] Given these circumstances, it is not in Egypt's interests to promulgate results that suggest Nile flows – that is, Egypt's main water supply – will be higher in the future. Rather than supporting Egypt's negotiating position, such results could be used by upstream states to justify expansion of their own water use. The sensitivity of this topic came up in a number of my interviews. When speaking to one non-Egyptian scientist who has published a paper in which most of the model runs presented indicate a future increase in Nile flows, I asked him how ministry officials responded to his data. His reply was revealing. 'There were a lot of political sensitivities around this', he said. 'That's another dimension of the work. I think it's the kind of result that isn't discussed much. If it's wetter in Egypt, it's all very uncertain'. In other words, when it is a matter of increased flows, ministry officials emphasize the uncertainty. A similar picture was given by an Egyptian scientist who had worked on a downscaling project. 'The range we ended up with was much smaller and it was actually more on the positive side than the negative side', he told me. I asked him how ministry officials responded. 'We had mixed responses,' he said. 'Some of them understand the issue and then they questioned the reliability of the results and how certain [we are] about the results. Also they questioned that the work we did was only based on one scenario of climate change'. In light of results that did not produce a politically convenient future, therefore, the officials utilized the uncertainty to undermine those results.

Uncertainty here is politically productive (Hastrup 2013; Limbert 2008; Mathews 2014; Taddei 2012; 2013). It offers a fuzzy space in which political possibilities emerge; a space in which scientists and policy-makers using the science may selectively interpret the future to reinforce Egypt's bargaining hand in its negotiations over shared use of the Nile. This manoeuvring to utilize the uncertainty in a politically expedient manner was explicit in one interview with an Egyptian scientist. 'Some [models] used to say it [Nile flow] will increase, some others to say it will decrease. Recently, I have heard more in the direction of increase, but I cannot say a definite word on this because of the uncertainty'. He concluded, 'I think there will always be some models saying reduction ... The last work we did ... most of it was increase. But we cannot eliminate the possibility of a decrease'. Even though their models suggest one thing, the scientist uses uncertainty to justify the necessity of considering the opposite.

The tension between what the models are saying and the government's imperative to maintain a picture of scarcity to bolster its claim on Nile waters is evident in the recent climate change adaptation strategy produced for the Ministry of Water Resources and Irrigation. The report is written by a water resource scientist who has a Ph.D. in civil engineering and heads one of the ministry's research institutes. In the report, he acknowledges,

> There is some disagreement between the general circulation models (GCMs) about whether the Blue Nile, which contributes more than 75% of Nile flows, will get wetter or drier. Roughly two thirds of the GCMs project an increase in precipitation in East Africa, while the others expect reduced precipitation. This means that the future of climate change in East Africa is uncertain (El-Din 2013: 36).

So his concluding statement is one of uncertainty. Unlike the non-Egyptian scientist I cited above, he does not use the fact that the majority of the models project a wetter outcome to conclude that an increase in flows is more likely. Instead he goes on to note that it is 'virtually certain' that the Nile Basin will become warmer. This warming will affect water availability by increasing evapotranspiration, although the degree to which this will outweigh any potential increase in precipitation is a matter of debate. None the less, he concludes, 'this uncertainty should not paralyze policy makers and water managers and stop them from rethinking and reevaluating current policies ... Any decrease in the total supply of water, coupled with an expected increase in consumption, would have drastic impacts' (El-Din 2013: 36). The paragraph thereby makes an interesting transition. It starts by noting how most GCMs indicate that East African precipitation is likely to increase, leading to increased Nile discharge. By the end

Journal of the Royal Anthropological Institute (N.S.), 46-66
© Royal Anthropological Institute 2016

of the paragraph, however, the author makes a conclusion pertaining only to the other scenario: that of a drier basin and increasing scarcity. In doing so, he subtly reinforces concerns about scarcity, in effect erasing the possibility of an increase in flows.

In these moments of encounter between scientists and non-scientists, therefore, whether in an interview with an anthropologist, a presentation to an international conference, a peer-review publication, or a policy report, we see the different ways in which scientists deal with the uncertainty bandwidth. While scientists have a complex language of probabilities to talk about the range of model predictions with one another, when speaking with non-experts, we see a notable variability in how they present that range. In bringing to the fore the models that indicate a negative impact on Nile flows from climate change, Egyptian scientists are able to reinforce broader national imperatives of sustaining the country's water supply from the Nile and mitigating against scarcity.

Signal in the uncertainty

All scientists negotiate complex positions, priorities, and perspectives in the process of producing knowledge. By looking in detail, though, at how one particular group of scientists produces and interprets knowledge about the future of a critical resource, in contrast with non-national counterparts, this paper sheds light on how one dimension of this knowledge – uncertainty – may be mobilized or underplayed for a variety of reasons.

The paper brings together three threads that highlight the relationship between scientists' knowing of uncertainty and their positioning within various assemblages. While the discussion is specific to Egypt, the paper's central argument – that scientists' engagements with uncertainty are shaped by their material, cultural, and political positionalities – speaks to wider anthropological debates on scientific knowledge-making. First, there is the uncertainty about how much rain there will be over East Africa in coming decades. Given that it is this rainfall that runs over the highlands and feeds the Nile, such knowledge is vital to understanding more about the river's future. Indeed these precipitation data sourced from GCMs are the starting-point for researching the impact of climate change on the Nile. Since the Egyptian scientists discussed here, by virtue of their employment and, in many cases, their research interests, are committed to examining the impacts of climate change on the Nile, it is not surprising that they, contrary to many non-Egyptian scientists, endorse the quality of the GCM precipitation data, despite the uncertainty of this information. Thus any scientist's position within national expert networks is significant for shaping how he or she engages with particular types of uncertainty.

The second thread shifts the focus to look at the relationship between uncertainty and material things. One of the ways in which scientists manage the uncertainties of climate modelling is by doing multiple simulations with different models, data, and parameters, and comparing the results. This process is contingent, though, on the materials needed to obtain, store, process, and analyse data. A scientist's ability to know the future with a greater degree of certainty depends, therefore, in part on his or her position within the international scientific community and the access to particular materials that that position affords.

The final part of the paper looks at the ultimate output of the model cascade: the projections of Nile flows. These projections are closely tied both to the GCM precipitation data (which are the main determinant of river flow) and to the materially

Journal of the Royal Anthropological Institute (N.S.), 46-66
© Royal Anthropological Institute 2016

mediated conduct of multiple simulations (which determines the number and spread of results on which those projections are based). While modelling studies suggest, statistically speaking, that there is a greater likelihood of Nile flows increasing in coming decades, Egyptian scientists' conclusion that Egypt's water supply is likely to decrease can be understood in light of their country's position within the Nile Basin, and the widespread concern within Egypt about challenges to its claim on Nile waters. Uncertainty here is linked to a scientist's position within the broader geopolitical context of a shared river basin.

The term that scientists use to describe an output of a modelling process that indicates either an increasing or decreasing trend is 'uncertainty in the signal'. In this paper, the signals at work are not only the positive or negative sign in front of the percentage change in Nile flows several decades from now. They are also the signals that a scientist makes to an international scientific community (including me, as a non-Egyptian researcher) to assert the validity of his or her modelling work; the signals that lie not in a human's ability to conduct a particular action, but in the materials that mediate those actions; and the signals that a downstream nation makes to upstream neighbours about the severity of its water outlook and the justice of its claims on shared waters. Hence while there is no doubt uncertainty in the signal, there is also much signalling in the uncertainty.

NOTES

I thank Patrick Gallagher, Evan Killick, David Kneas, Sarah Laborde, Myanna Lahsen, Andrew Mathews, Jessica O'Reilly, and the anonymous reviewers for their insightful comments on drafts of this paper. I also thank Greg Carbone, Kirstin Dow, and Aashka Patel for helping me understand more about climate modelling.

[1] Temperatures are also shifting under climate change, and thus the loss of water from the Nile Basin through evapotranspiration is also likely to change, but studies have suggested that this impact will be secondary (Conway & Hulme 1996; Sayed 2004; Sene, Tate & Farquharson 2001).

[2] During my fieldwork with farmers and irrigation engineers, I did not find the future of the Nile to be a prominent topic of conversation. Farmers and engineers talked about the variability in the supply of Nile water at a local level and many expressed concern about scarcity. But the Nile itself, Egypt's water supply, was not seen in temporal terms as a declining resource. (At this time, in 2007-8, Ethiopia had yet to start building the dam across the Nile's main tributary.) In contrast with non-renewable resources like oil, where the end of the resource is always on the horizon and whose future haunts the present (Limbert 2010), the future of the Nile does not infuse everyday life. I also did not find climate change to be a topic of everyday conversation among Egyptian farmers and irrigation engineers. (This is a point of contrast with other studies that have highlighted local perceptions of climate change: Crate & Nuttall 2009; Orlove, Chiang & Cane 2002; Rudiak-Gould 2012.) The reason for the absence of talk of climate change in daily life is perhaps because the main change in climate that will affect the water supply from the Nile – precipitation – is occurring not in Egypt, where there is very little rainfall, but thousands of kilometres away in the East African Highlands. The climate that people in rural Egypt experience, therefore, is disconnected from the scarcity or abundance of their main source of water (Barnes 2015).

[3] The degree to which scientific knowledge about Nile futures does actually translate into contemporary changes in water management policy within Egypt is, however, another matter, beyond the scope of this paper's analysis.

[4] I use the term 'scientist' to refer to people studying climate change, without distinguishing between modellers and model users (groups that in the case of regional modelling, are not so distinct). In Egypt, these people are generally referred to as 'engineers' – if their first degree was in engineering – or 'doctors', if they have Ph.D.s. I follow this format in the pseudonyms that I assign all my informants.

[5] All these data were collected in English – the principal language of climate science and the dominant language at international water conferences. Owing to the nature of these data, the paper does not speak to the question of how scientists talk among themselves when in non-public settings, such as in the computer lab. It also does not delve into Egyptian decision-makers' perspectives and how they might diverge from those of national scientists (e.g. Lahsen 2009).

[6] Many scientists are careful to distinguish between prediction and projection. They reject the former term owing to its implications of a certainty that is not in fact present. In practice, however, the two terms are often used interchangeably, as I do in this paper.

[7] The GCMs currently in use have been developed and are being run by modelling groups in Western Europe, North America, Japan, China, Russia, Australia, and South Korea.

[8] The process of downscaling may also be used to obtain data that are more temporally detailed.

[9] Downscaling can also be done through statistical methods, as in the case of some of the earlier studies on the Nile (e.g. Elshamy, Sayed & Badawy 2009).

[10] For a fascinating account of the institutional and discursive associations that allow the Hadley Centre's regional climate model to move to new sites of climate simulation in the Global South, see Mahony & Hulme (2012).

[11] Notably, there may be a difference between how scientists present their level of confidence in the data to me and how they talk about it to each other.

[12] This scientist's shift in his research focus may also have been an opportunistic response to the fact that adaptation has become a priority area for research funding.

[13] A similar level of commitment to working on projects that reflect and respond to national interests is noted by Mahony (2014) in his work with climate scientists in government research institutes in India. This link between scientists' national identity and their work has also been underscored by the work on nationalism and science (e.g. Abraham 1998).

[14] A similar theme emerged in the Arab Forum when an Egyptian scientist gave a presentation arguing that the Arab countries should invest in developing their own GCM so that they are not reliant on GCM data produced by others.

[15] The most recent IPCC report has replaced the emissions scenarios with a method of projecting emissions according to 'representative concentration pathways'.

[16] There are a number of ongoing international initiatives to facilitate broader access to GCM data (e.g. the Coupled Model Intercomparison Project), but none of the Egyptian scientists whom I interviewed had begun to tap into these data sources.

[17] The mean download speed in Egypt is 1.3 megabits per second, whereas in the United States it is 25.7 megabits per second. Data from testmy.net, last accessed on 11 January 2016.

[18] While many scientists make statements like this, whether or not it is statistically true to say that a certain outcome is more likely because more models produced that outcome depends on certain assumptions, such as that each model run is independent of the others (Knutti 2008).

[19] The scientist's qualification that this result is for the next few decades is significant. A number of studies have indicated an increase in Nile flows over coming decades (e.g. through 2040), but then a decrease after that (e.g. from 2040 to 2100 in Beyene *et al.* 2010). This complexity is often lost in translation, though, when the outlook for the Nile becomes collapsed into an undifferentiated future, rather than one that varies depending on the decades in question.

[20] In March 2015, the leaders of Ethiopia, Sudan, and Egypt signed a declaration of principles agreement on Ethiopia's Grand Renaissance Dam. Whether this preliminary agreement evolves into a binding regional agreement that Egypt supports, however, remains to be seen.

REFERENCES

ABRAHAM, I. 1998. *The making of the Indian atomic bomb*. London: Zed Books.

APPADURAI, A. 1998. Dead certainty: ethnic violence in the era of globalization. *Development and Change* **29**, 905-25.

BARNES, J. 2014. *Cultivating the Nile: the everyday politics of water in Egypt*. Durham, N.C.: Duke University Press.

——— 2015. Scale and agency: climate change and the future of Egypt's water. In *Climate cultures: anthropological perspectives on climate change* (eds) J. Barnes & M. Dove, 127-45. New Haven: Yale University Press.

BEYENE, T., D. LETTENMAIER & P. KABAT 2010. Hydrologic impacts of climate change on the Nile River Basin: implications of the 2007 IPCC scenarios. *Climatic Change* **100**, 433-61.

BHATTACHARJEE, P. & B. ZAITCHIK 2015. Perspectives on CMIP5 model performance in the Nile headwaters regions. *International Journal of Climatology* **35**, 4262-75.

BLOSCHL, G. & A. MONTANARI 2010. Climate change impacts – throwing the dice? *Hydrological Processes* **24**, 374-81.

Journal of the Royal Anthropological Institute (N.S.), 46-66

BOHOLM, A. 2003. The cultural nature of risk: can there be an anthropology of uncertainty? *Ethnos: Journal of Anthropology* **68**, 159-78.

BRYSSE, K., N. ORESKES, J. O'REILLY & M. OPPENHEIMER 2013. Climate change prediction: erring on the side of least drama? *Global Environmental Change* **23**, 327-37.

CADUFF, C. 2014. Prophetic prophecy, or how to have faith in reason. *Current Anthropology* **55**, 296-315.

CASCÃO, A. 2008. Ethiopia – challenges to Egyptian hegemony in the Nile Basin. *Water Policy* **10**, 13-28.

———— 2009. Changing power relations in the Nile River Basin: unilateralism vs cooperation? *Water Alternatives* **2**, 245-68.

COLLINS, R.O. 1994. History, hydropolitics and the Nile: Nile control – myth or reality? In *The Nile – sharing a scarce resource: an historical and technical review of water management and of economical and legal issues* (eds) P. Howell & J. Allan, 109-37. Cambridge: University Press.

CONWAY, D. 2005. From headwater tributaries to international river: observing and adapting to climate variability and change in the Nile Basin. *Global Environmental Change* **15**, 99-114.

———— & M. HULME 1993. Recent fluctuations in precipitation and runoff over the Nile sub-basins and their impact on main Nile discharge. *Climatic Change* **25**, 127-51.

———— & ———— 1996. The impacts of climate variability and future climate change in the Nile Basin on water resources in Egypt. *International Journal of Water Resources Development* **12**, 261-80.

CRATE, S. & M. NUTTALL (eds) 2009. *Anthropology and climate change*. Walnut Creek, Calif.: Left Coast Press.

CURRY, J. & P. WEBSTER 2011. Climate science and the uncertainty monster. *Bulletin of the American Meteorological Society* **92**, 1667-82.

DOUGLAS, M. & A. WILDAVSKY 1983. *Risk and culture: an essay on the selection of technical and environmental dangers*. Berkeley: University of California Press.

EDWARDS, P.N. 2010. *A vast machine: computer models, climate data, and the politics of global warming*. Cambridge, Mass.: MIT Press.

EL-DIN, M. 2013. Proposed climate change adaptation strategy for the Ministry of Water Resources and Irrigation. Prepared for the UNESCO-Cairo Office, Joint Programme for Climate Change Risk Management in Egypt, January.

ELSHAMY, M., M. SAYED & B. BADAWY 2009. Impacts of climate change on the Nile flows at Dongola using statistical downscaled GCM scenarios. *Nile Basin Water Engineering Scientific Magazine* **2**, 1-14.

————, I. SEIERSTAD & A. SORTEBERG 2009. Impacts of climate change on Blue Nile flows using bias-corrected GCM scenarios. *Hydrology and Earth System Sciences* **13**, 551-65.

GLEICK, P. 1991. The vulnerability of runoff in the Nile Basin to climatic changes. *The Environmental Professional* **13**, 66-73.

GUSTERSON, H. 1998. *Nuclear rites: a weapons laboratory at the end of the Cold War*. Berkeley: University of California Press.

HASTRUP, K. 2013. Anticipating nature: the productive uncertainty of climate models. In *The social life of climate models: anticipating nature* (eds) K. Hastrup & M. Skrydstrup, 1-29. New York: Routledge.

HAWKINS, E. & R. SUTTON 2009. The potential to narrow uncertainty in regional climate predictions. *Bulletin of the American Meteorological Society* **90**, 1095-107.

———— & ———— 2011. The potential to narrow uncertainty in projections of regional precipitation change. *Climate Dynamics* **37**, 407-18.

HELMREICH, S. 2009. *Alien ocean: anthropological voyages in microbial seas*. Berkeley: University of California Press.

JOHNSON-HANKS, J. 2005. When the future decides: uncertainty and intentional action in contemporary Cameroon. *Current Anthropology* **46**, 363-85.

KINGSTON, D. & R. TAYLOR 2010. Sources of uncertainty in climate change impacts on river discharge and groundwater in the headwater catchment of the Upper Nile Basin, Uganda. *Hydrology and Earth System Sciences* **14**, 1297-308.

KNORR CETINA, K. 1999. *Epistemic cultures: how the sciences make knowledge*. Cambridge, Mass.: Harvard University Press.

KNUTTI, R. 2008. Should we believe model predictions of future climate change? *Philosophical Transactions of the Royal Society A* **366**, 4647-64.

KUNDZEWICZ, Z. & E. STAKHIV 2010. Are climate models 'ready for prime time' in water resources management application, or is more research needed? *Hydrological Sciences Journal* **55**, 1085-9.

LAHSEN, M. 2005. Seductive simulations: uncertainty distribution around climate models. *Social Studies of Science* **35**, 895-922.

Journal of the Royal Anthropological Institute (N.S.), 46-66
© Royal Anthropological Institute 2016

————— 2007. Trust through participation? Problems of knowledge in climate decision making. In *Social construction of climate change: power, knowledge, norms, discourse* (ed.) M. Pettenger, 173-96. London: Ashgate.

————— 2009. A science-policy interface in the Global South: the politics of carbon sinks and science in Brazil. *Climatic Change* **97**, 339-72.

LAKOFF, A. 2008. The generic biothreat, or, how we became unprepared. *Cultural Anthropology* **23**, 399-428.

LATOUR, B. 1988. *Science in action: how to follow scientists and engineers through society*. Cambridge, Mass.: Harvard University Press.

————— & S. WOOLGAR (eds) 1986. *Laboratory life* (Second edition). Princeton: University Press.

LIMBERT, M. 2008. Depleted futures: anticipating the end of oil in Oman. In *Timely assets: the politics of resources and their temporalities* (eds) E. Ferry & M. Limbert, 25-50. Santa Fe, N.M.: School for Advanced Research Press.

————— 2010. *In the time of oil: piety, memory and social life in an Omani town*. Stanford: University Press.

MAHONY, M. 2014. The predictive state: science, territory and the future of the Indian climate. *Social Studies of Science* **44**, 109-33.

————— & M. HULME 2012. Model migrations: mobility and boundary crossings in regional climate prediction. *Transactions of the Institute of British Geographers* **37**: 2, 197-211.

MATHEWS, A. 2014. Scandals, audit, and fictions: linking climate change to Mexican forests. *Social Studies of Science* **44**, 82-108.

MEARNS, L. 2010. The drama of uncertainty. *Climatic Change* **100**, 77-85.

MEKONNEN, D. 2010. The Nile Basin cooperative framework agreement negotiations and the adoption of a 'water security' paradigm: flight into obscurity or a logical cul-de-sac? *European Journal of International Law* **21**, 421-40.

MURPHY, M. 2006. *Sick building syndrome and the problem of uncertainty*. Durham, N.C.: Duke University Press.

NAWAZ, R., T. BELLERBY, M. SAYED & M. ELSHAMY 2010. Blue Nile runoff sensitivity to climate change. *The Open Hydrology Journal* **4**, 137-51.

NICOL, A. & A. CASCÃO 2011. Against the flow – new power dynamics and upstream mobilisation in the Nile Basin. *Review of African Political Economy* **38**, 317-25.

O'REILLY, J., K. BRYSSE, M. OPPENHEIMER & N. ORESKES 2011. Characterizing uncertainty in expert assessments: ozone depletion and the West Antarctic Ice Sheet. *WIREs Climate Change* **2**, 728-43.

ORLOVE, B., J. CHIANG & M. CANE 2002. Ethnoclimatology in the Andes. *American Scientist* **90**, 428-35.

PETRYNA, A. 2002. *Life exposed: biological citizens after Chernobyl*. Princeton: University Press.

RAPP, R. 2000. *Testing women, testing the foetus: the social impact of amniocentesis in America*. New York: Routledge.

RAVETZ, J. 1987. Usable knowledge, usable ignorance: incomplete science with policy implications. *Science Communication* **9**, 87-116.

REDFIELD, P. 2013. *Life in crisis: the ethical journey of Doctors Without Borders*. Berkeley: University of California Press.

RUDIAK-GOULD, P. 2012. Promiscuous corroboration and climate change translation: a case study from the Marshall Islands. *Global Environmental Change* **22**, 46-54.

SAMIMIAN-DARASH, L. 2013. Governing future potential biothreats: toward an anthropology of uncertainty. *Current Anthropology* **54**, 1-22.

SAREWITZ, D., R. PIELKE & R. BYERLY (eds) 2000. *Prediction: science, decision making, and the future of nature*. Washington, D.C.: Island Press.

SAYED, M. 2004. Impacts of climate change on Nile flows. Ph.D. dissertation, Ain Shams University, Cairo.

SENE, K., E. TATE & F. FARQUHARSON 2001. Sensitivity studies of the impacts of climate change on White Nile flows. *Climatic Change* **50**, 177-208.

SHACKLEY, S. & B. WYNNE 1996. Representing uncertainty in global climate change science and policy. *Science, Technology, and Human Values* **21**, 275-302.

SOLIMAN, E., A. SAYED & M. JEULAND 2009. Impact assessment of future climate change for the Blue Nile Basin, using a RCM nested in a GCM. *Nile Basin Water Engineering Scientific Magazine* **2**, 15-30.

STAR, S. 1985. Scientific work and uncertainty. *Social Studies of Science* **15**, 391-427.

STRZEPEK, K., D. YATES & D. EL-QUOSY 1996. Vulnerability assessment of water resources in Egypt to climatic change in the Nile Basin. *Climate Research* **6**, 89-95.

TADDEI, R. 2012. The politics of uncertainty and the fate of forecasters. *Ethics, Policy and Environment* **15**, 252-67.

———— 2013. Anthropologies of the future: on the social performativity of (climate) forecasts. In *Environmental anthropology: future directions* (eds) H. Kopnina & E. Shoreman-Ouimet, 244-63. London: Routledge.

TAYE, M., V. NTEGEKA, N. OGIRAMOI & P. WILLEMS 2011. Assessment of climate change impact on hydrological extremes in two source regions of the Nile River Basin. *Hydrology and Earth System Sciences* **15**, 209-21.

TRAWEEK, S. 1992. *Beamtimes and lifetimes: the world of high energy physicists.* Cambridge, Mass.: Harvard University Press.

VAN DER STEEN, M. 2008. Ageing or silvering? Political debate about ageing in the Netherlands. *Science and Public Policy* **35**, 575-83.

WATERBURY, J. 1979. *Hydropolitics of the Nile Valley.* Syracuse: University Press.

———— 2002. *The Nile Basin: national determinants of collective action.* New Haven: Yale University Press.

WRIGHT, G. & L. PHILLIPS 1979. Cross-cultural differences in the assessment and communication of uncertainty. *Current Anthropology* **20**, 845-6.

YATES, D. & K. STRZEPEK 1998. Modeling the Nile Basin under climatic change. *Journal of Hydrologic Engineering* **3**, 98-108.

L'incertitude du signal : modéliser les futurs hydrologiques de l'Égypte

Résumé

L'impact du changement climatique sur le Nil est lourd de conséquences pour l'Égypte, qui tire presque toute son eau du fleuve. Pourtant, les modèles utilisés pour le prédire donnent des résultats très discordants, de prévisions annonçant une augmentation du débit à des pronostics inverses. Cette incertitude occupe une place plus ou moins centrale dans les débats sur l'impact du changement climatique dans le monde et peut être exploitée à diverses fins politiques. Bien que tous les scientifiques négocient l'incertitude dans leur pratique quotidienne, l'auteure se concentre ici sur deux groupes précis : les chercheurs égyptiens et étrangers qui travaillent sur la modélisation des futurs du Nil. Pour commencer, elle examine la manière dont les égyptiens minimisent des incertitudes essentielles, en relation avec leur position au sein des réseaux nationaux d'experts qui prédisent l'avenir hydrologique de leur pays. Ensuite, elle met en lumière une dimension fondamentale de la capacité qu'ont les scientifiques à reconnaître, affronter et réduire l'incertitude. Enfin, elle avance que l'interprétation sélective des résultats de modélisation incertains par les scientifiques égyptiens, qui se concentrent sur ceux qui prédisent une baisse du débit, s'inscrit dans une tentative politique de conforter les revendications de l'égypte sur les eaux du Nil. En explorant l'incertitude comme une entité connue que les scientifiques gèrent activement, non seulement en égypte mais dans tous les contextes politiques, cet article met en lumière les liens entre l'incertitude du savoir sur différents futurs et les positions matérielles, culturelles et politiques de ceux qui produisent ce savoir.

Journal of the Royal Anthropological Institute (N.S.), 46-66
© Royal Anthropological Institute 2016

4

Subsoil abundance and surface absence: a junior mining company and its performance of prognosis in northwestern Ecuador

DAVID KNEAS *University of South Carolina*

Spectacle and performance have long characterized global mining investment. In this paper, I examine the performative strategies of a junior mining company called Ascendant Copper in its pursuit of copper in northwestern Ecuador in the mid-2000s. I focus on Ascendant's contradictory efforts to attract mining investment with imagery of subsoil abundance while also downplaying the scale and significance of mining in the face of local opposition. I approach the company's divergent strategies as reflective of the dual meanings of prognosis, as both forecast and diagnosis. I begin by examining the means through which Ascendant promoted subsoil copper wealth within the framework of resource classification established by the Toronto Stock Exchange. I then analyse the way the company employed discourses of corporate social responsibility to portray its local presence as one defined by conservation and development. In analysing the articulation of a junior company like Ascendant in relation to both subsoil and surface, I not only highlight the materiality of corporate social responsibility, but also underscore the precarious becoming of junior companies, who seldom feature in anthropological accounts of corporations.

In the early 1980s, a team of Ecuadorian and Belgian geologists identified a 'strong copper anomaly' in the Intag region of northwestern Ecuador, in the watershed above a remote community called Junin (Piedra & Arrata 1987: 12). Subsequent sediment surveys indicated the potential for Junin to be what geologists described then as 'a first-order mineral prospect' (DGGM 1985: 3). In conjunction with neoliberal reforms to Ecuador's mining laws, Bishimetals (a subsidiary of the Japanese conglomerate Mitsubishi) began subsoil explorations in 1991. The company drilled an average of seven exploratory drill holes per year in the watershed above Junin. This exploration came to a halt in May 1997, however, when farmers from in and around Junin occupied and burned down Bishimetals' mining camp. In its place, they left a sign that said miners would never again set foot in the community.

No miner did until 2004, when a Canadian company called Ascendant Copper acquired the rights to the subsoil concession. Ascendant was a 'junior'. Within the global mining industry, junior companies are those whose primary interest is resource exploration, particularly of so-called 'greenfield' sites: areas with little or no prior history of mineral extraction, like Intag and, to a certain extent, Ecuador as a whole. Like most juniors, Ascendant depended upon the financial investment it could garner on the Toronto Stock Exchange. Yet, owing to a roadblock erected by local opposition in Junín, the company was unable to access the concession and begin the process of resource exploration. It was therefore forced to rely on rock samples extracted by Bishimetals. Based on those same samples, Ascendant forecast a subsoil resource over four times that estimated by the Japanese company. This estimate became central to Ascendant's claims of resource potential. At the same time, however, in its interactions with local communities, Ascendant downplayed the prospect and scale of mining for Intag's future. Instead, it sold itself as an organization of conservation and development whose primary interest was in identifying and solving what it defined as the region's associated problems of poverty and environmental degradation. In this paper, I analyse the motivations behind and the contradictions within this paradox of subsoil abundance and surface absence. As I describe here, Ascendant attempted to navigate this contradiction through a politics of prognosis that reflected the dual meanings of the term.

Prognosis means both forecast and diagnosis, the language of weather prediction *and* medical practice. In relation to potential investors, Ascendant articulated a discourse of *prognosis as prediction* to describe a resource that would become one of the world's largest copper mines. In local contexts and more public venues, by contrast, the company spoke a language of *prognosis as diagnosis* in its portrayal of Intag's surface landscape as degraded, threatened, and in need of outside intervention. Accounts of potential subsoil abundance, produced within and against notions of geological knowledge, are, at the same time, forecasts of mineral futures. The language of diagnoses, on the other hand, is an idiom of expertise that focuses attention on the immediate present, and the near futures that may arise with or without intervention. Both iterations of prognosis, then, reflect contrasting stories of becoming, of a region, a mineral deposit, and a mining company. In this sense, the dialectics of prognosis I explore here are not unique to Ascendant Copper, but reflect the constraints and temporalities that, to varying degrees, confront all projects of mineral exploration, especially those in 'greenfield' sites.

I approach Ascendant's discourse of subsoil and surface prognosis through a framework of performance. Performance, according to Schieffelin, is 'concerned with something that anthropologists have always found hard to characterize theoretically: the creation of presence' (1998: 194). Presence, of course, highlights the anxieties within Ascendant's divergent forms of prognosis: the presence of subsoil copper and the presence, both short and long term, of Ascendant Copper, in the Intag region and as a viable mining company. As Edward Schieffelin notes, performance relates to presence not as existence, but rather as creation. Performance highlights process. It helps address the means, intentions, and forms of knowledge through which Ascendant portrayed subsoil copper and surface terrains, and the company's relation to both. Performativity, according to Karen Barad, 'shifts the focus from questions of correspondence between descriptions and reality to matters of practices, doings, actions' (2003: 802).

Recent scholarship on resource materiality has likewise emphasized matters of practice in examining how specific resources come into being (Bakker & Bridge 2006; Bridge 2009; Richardson & Weszkalnys 2014). A notion of becoming, in this sense, does not assume the *a priori* existence of subsoil resources, but rather calls attention to the means through which natural resources are conjured as objects of significance. In his discussion of the divides between material culture and ecological anthropology,

Journal of the Royal Anthropological Institute (N.S.), 67-86
© Royal Anthropological Institute 2016

Tim Ingold also argues for a more phenomenological understanding of materials, of studying the process and practice through which situated actors 'draw out and bring forth potentials immanent in a world of becoming' (2012: 435). Employing Barad's notions of performativity, he argues that 'to describe any material is to pose a riddle' that must be answered by following it, of situating the 'movements and gestures' that go into the manufacturing of materiality; or in the case of this paper, resource potential (Ingold 2012: 435).

In the first part of the paper, I compare the resource estimates put forward by Bishimetals and Ascendant Copper. Inspired by Ingold, I am interested in a particular riddle: through what means, and towards what ends, did Ascendant produce an estimate of subsoil copper four times larger than Bishimetals, even though the two companies used the same mineralogical data? In answering this question, I examine rather mundane techniques of 'appraisal and human labor' (Richardson & Weszkalnys 2014: 12). Specifically, I focus on different techniques of block modelling and search radius, as well as the epistemological field that authorized these differences. By this I mean the rules of resource classification established by the Toronto Stock Exchange in 2000. The Toronto rules emerged in response to Bre-X, a junior mining company whose fraudulent claims to subsoil abundance unsettled the edifice of global mining investment in 1997. Anna Tsing (2000) argues that, far from being a complete outlier, the Bre-X fraud is revelatory of the dramaturgy that defines global mining investment. Tsing writes that Bre-X 'was always a performance, a drama, a conjuring trick, an illusion, regardless of whether real gold or only dreams of gold ever existed' (2000: 118). In this sense, I approach Ascendant's resource estimate not as a thing, but as a doing, an ensemble of performative actions that gesture towards the prospect of abundant subsoil copper.

In the second part of the paper, I analyze Ascendant's above-ground prognosis (as diagnosis) in relation to discourses of corporate betterment that have emerged since the millennium under the distinct field and banner of corporate social responsibility (CSR). Diagnostic activity, geared towards identifying and categorizing problems and their potential solutions (Buscher, Goodwin & Mesman 2010: 1), epitomizes the way scholars have analysed the politics of CSR (Benson 2008; Billo 2015; Thieme, Bedi & Vira 2015). For the extractive industries in particular, mining companies have increasingly drawn on ideas and practices of CSR to better dictate the terms of engagement with local communities, environmental non-governmental organizations, and nation-states (Jenkins 2004; Kirsch 2014). While corporate leaders describe CSR as embodying the ultimate win-win, many critics see it as nothing more than a veil to protect corporate profits. For Peter Benson and Stuart Kirsch, CSR is the ultimate corporate oxymoron: the linguistic appraisal of contradictory terms, like *sustainable mining*, that corporations employ to 'promote doubt or ambiguity concerning the harms they cause' (Benson & Kirsch 2010: 47).

Whereas critics like these view CSR through the prism of public relations, other scholars have used theories of performativity to describe the cultural processes that constitute the core tenets of corporate responsibility. In her ethnographic study of global CSR meetings and policy forums, Dinah Rajak examines the 'highly ritualistic theatres of virtue' (2011*b*: 11) that comprise the global arena of CSR and the norms and worldviews that sustain it. Rajak (2010; 2011*a*) views CSR as more than a form of image management, and approaches it as a dynamic field of corporate practice and knowledge. In her recent ethnography of Newmont mining, Marina Welker (2014)

Journal of the Royal Anthropological Institute (N.S.), 67-86
© Royal Anthropological Institute 2016

echoes these points, but also criticizes the image of a uniform, profit-driven corporate monolith that she sees as lurking in many of the critical assessments of what CSR is and what it accomplishes. In this regard, Welker emphasizes corporate *enactments*, the different ways various actors, from corporate managers, environmental activists, to local villagers, imagine the corporation, what it does, and what they can expect from it. As with research on resource materialties, Welker shifts attention away from *a priori* corporate beings to questions of indeterminate corporate becomings.

The recent monographs by Rajak (2011*a*) and Welker (2014), focused on mining giants Anglo-American and Newmont, are both pioneering ethnographies. In their movement between global headquarters and on-the-ground mining sites, they are quintessentially multi-sited. Despite their interest in corporate enactments and the sustaining power of CSR, however, Welker and Rajak do not subject platinum and gold, as subsoil resources and global commodities, to the same analytical force as they do other dimensions of corporate behaviour and self-fashioning. Nevertheless, these scholars raise fundamental questions about the social lives of both corporations and CSR, questions that a focus on a company like Ascendant Copper is particularly helpful at illuminating. Part of what makes a junior company 'junior' is its material contingency. Junior companies are precarious entities that live in a world of potential. Ascendant Copper's existence was nothing but a linked set of subsoil and surface enactments.

In this regard, even as theories of performance have informed recent scholarship on both natural resources and corporate social behaviour, performance assumes a level of mastery that does not square with Ascendant's more junior, and indeed ephemeral, status. A more apt framing might be the theatre workshop. Workshops, according to performance theorist Richard Schechner (1985), are the most fluid yet unstable settings in theatre production. They are the creative space where actors try out combinations of behaviour that would not be possible in the stricter confines of rehearsals or performances. As opposed to refined actors and defined spectators, workshop performances are more raw and more in the realm of fantasy. 'The fantasies of workshop help the performer in rehearsal "find" the character who during performance "presents" the narrative to the spectator' (Schechner 1985: 291). Whereas performance carries with it notions of finality, the theoretical space of the workshop underscores the temporalities and the differential fields of power that characterize the world of junior mining companies like Ascendant.

Ascendant Copper no longer exists. Despite the company's predictions of copper wealth and its self-proclaimed aim to 'heal, cultivate, and grow the people and land of Ecuador' (AscendantCopper.com n.d.), by 2009 it was coming undone. The social life of Ascendant (2004-9) corresponds to an intense period of mining investment and exploration in South America (Bury & Bebbington 2013). This post-millennial boom was both a continuation of investment that began with sector-wide neoliberal restructuring in the early 1990s and a recovery from the Bre-X scandal of 1997 (Bridge 2009). It also marked a high point for junior mining companies and the consolidation of the Toronto Stock Exchange as the global centre of mining investment (Fulp 2009; Lowell 2001; Trevethan 2005).

In examining Ascendant's politics of prognosis in relation to both CSR and the Toronto rules of resource classification – two defining fields of post-millennial corporate practice – I present an important addition to recent work on the varied dimensions of mineral exploration and mining conflict within Ecuador, the Andes, and South America as a whole (Bebbington 2012*b*; Bebbington & Bury 2013).[1] At the same time,

Journal of the Royal Anthropological Institute (N.S.), 67-86
© Royal Anthropological Institute 2016

I also further discussion on resource temporalities (Ferry & Limbert 2008) and the materiality of corporate social practice (Rogers 2012; Rolston 2013). The latter body of literature is itself emblematic of recent interest in the social lives and modalities of corporations (Appel 2012; Welker, Partridge & Hardin 2011).

This paper draws on historical and ethnographic research that I have carried out in Intag and Ecuador since 2004. The first part of the paper is based on a formal audit of Ascendant's resource estimate and corresponding press releases that the Toronto Stock Exchange required as part of its post Bre-X framework of accountability. The second part of the paper is based on my own observations of Ascendant during research trips to Intag between 2004 and 2008 and is supported by an additional thirteen months of ethnographic fieldwork that I conducted between 2009 and 2010.[2] My assessment of Ascendant's discursive rendering of itself and its portrayal of Intag comes from close analysis of its assorted press releases, company reports, and various iterations of its website between 2004 and 2009. These websites, in which the company invested heavily, no longer exist on-line. My examination of these sites comes from my own archival collection that I have established over the past decade.

Prognosis as prediction

By the time it was expelled in 1997, Bishimetals had drilled thirty exploratory drill holes in the watershed above Junin. Each year the perforations went deeper, such that five of the last six holes descended 600 metres below the surface. Over six years of exploration, Bishimetals extracted almost 10,000 metres of rock core. Once extracted, the core material was divided into two sections, half of which was stored in wooden boxes to become a material archive, and half sent to laboratories in Quito for analysis of copper content (Micon International 2005). According to its final report issued in 1998, Bishimetals judged the Junin deposit to contain a total of 319 million tons of material, with a copper grade of 0.71 per cent copper (Metal Mining Agency of Japan 1998). Seven years later, Ascendant Copper estimated a total deposit of 982 million tons of rock, with a copper grade of 0.89 per cent (Ascendant Copper 2005a: 6). Ascendant, in essence, forecast a total copper resource over four times as large as that estimated by Bishimetals (Table 1).

Before outlining the specific means of Ascendant's resource appraisal, it is worth interpreting the underlying aims of the company's estimate. While Bishimetals' estimate of a 320 million ton deposit is nothing to scoff at, a total resource of 982 million tons, containing nearly 20 billion pounds of copper, would make Junin, in Ascendant's own words, 'one of the world's largest undeveloped copper deposits' (Copper Mesa 2009).[3] This type of statement, and the resource estimate that makes it possible, is precisely what Tsing describes as 'the self-conscious making of a spectacle' (2000: 118). A key part of the spectacle here is the way the 982 million ton estimate operates as an ordinal number (Guyer 2010), a ranking that links the Junin prospect with the pantheon of 'first order' global copper deposits, from Pebble Mine in Alaska to iconic copper mines in Chile. It is a spectacle that is even more necessary given Ascendant's lack of access to the mineral concession. Or, as an investment blogger called Deep_Digger put it in response to concerns over local opposition, 'that is a whole lot of copper no matter how you look at it'.[4] In other words, Ascendant was attempting to offset its lack of access – a red flag for investment risk – through the prospect of globally significant copper abundance. As Tsing writes in the case of Bre-X, 'No one would ever have invested in Bre-X if it had not created a performance, a dramatic exposition of the possibilities of

Table 1. Contrasting estimates of the Junin copper deposit

Source	Deposit size (million tons)	Copper grade (%)	Total copper (billion pounds)
Bishimetals (1998)	319	0.71	4.5
Ascendant (2005)	982	0.89	19

gold ... the self-conscious making of a spectacle [that] is a necessary aid to gathering investment funds' (2000: 118).

The underlying spectacle of Ascendant's inflated estimate highlights how the company's means of resource appraisal were geared towards a specific end of dramatic abundance. Ascendant produced this larger estimate by modifying some of the variables of subsoil appraisal that Bishimetals had employed in the late 1990s. Bishimetals analysed the Junin rock cores in 10-metre sections. It then utilized a software program called Minex and an estimate technique called block modelling to translate that data into the outlines of a porphyry copper deposit (Micon International 2005: 13). Block modelling is essentially a process whereby samples from the drill cores are projected to be large-tonnage subsoil blocks, which Bishimetals defined as 25 m in X and Y directions, and 10 m in the Z direction (Micon International 2005: 13). To define the mineral content of those blocks, Bishimetals used a search radius of 150 m, which is to say that sample points up to 150 m away defined the block characteristics. It is against these parameters and assumptions that Bishimetals derived a total resource of 319 million tons.

Interestingly, however, Bishimetals also stated that if a search radius of 50 m were used for block calculation – instead of 150 m – then the total resource size would decrease to 37 million tons (Micon International 2005: 14). This is not only an order of magnitude smaller, but a deposit that barely registers as an economically significant resource. In other words, the difference here between a rather modest copper resource of 37 million tons and a globally significant deposit of 982 million tons is the result not of geological processes but of simple modifications to techniques of block modelling and search radius.

Using a resource database that it assembled from Bishimetals, Ascendant used what it referred to in its reporting as 'proprietary software' to derive its own resource estimate (Micon International 2005: 35). Unlike Bishimetals, Ascendant utilized a larger block size ($X-75$ m, $Y-75$ m, and $Z-25$ m) and also incorporated a variable distance radius to calculate the contents of those blocks (Table 2). Block samples closer to the surface

Table 2. Ascendant and Bishimetals' methodologies of resource estimation

	Block size (m)	Distance radius used to calculate content of blocks (m)	Estimated size of Junin copper deposit (million tons)
Bishimetals' conservative estimate (1998)	25 × 25 × 10	50	37
Bishimetals' optimistic estimate (1998)	25 × 25 × 10	150	319
Ascendant estimate (2005)	75 × 75 × 35	Variable – larger at depth	982

Journal of the Royal Anthropological Institute (N.S.), 67-86
© Royal Anthropological Institute 2016

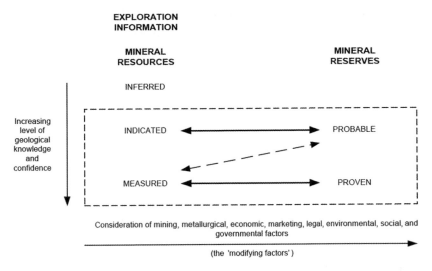

Figure 1. Canadian Institute of Mining categories of resource classification.

had a small radius, while blocks at the bottom of the deposit had a larger radius. What those radii were and how they varied is unclear, concealed within the term 'proprietary software'. But part of Ascendant's justification for the use of a variable radius was due to the fact that of the thirty total boreholes that Bishimetals drilled, only four reached a vertical depth of 600 metres. Ascendant assumed that the copper deposit was more abundant the deeper it went, a defining characteristic of first-order copper deposits. As Ascendant advertised in one press release, 'Logging shows that the intensity of cupriferous mineralization occurs at depth; in particular, four of the five holes that went to 600 meters all bottomed in intensely altered and mineralized porphyry copper' (Ascendant Copper 2006*a*). Despite the fact that Bishimetals had collected material from thirty boreholes, Ascendant placed special emphasis on the four holes that went to 600 metres, and did so precisely through the utilization of a larger block size and variable radius to define those blocks. The invocation of richer resource at depth, one evidenced by a few, yet promising samples, is just the sort of illusion of potential and possibility that Ascendant was trying to sell.

It was also the illusion that the Canadian authorities were keen to codify in the wake of the Bre-X fraud. In response to Bre-X, the Canadian Institute of Mining (CIM) created new standards for the disclosure of mining projects, a set of rules called National Instrument 43-101, a framework that resonates with what anthropologists have termed audit culture (Power 1997; Strathern 2000). In addition to mandating updated and timely press releases about company activities, the CIM rules also require listed companies to categorize resource estimates according to degrees of geological confidence. The CIM rules classify three types of mineral resources: 'inferred', 'indicated', and 'measured'. 'Indicated' and 'measured' are further differentiated by subcategories of 'probable' and 'proven' mineral reserves (Fig. 1). According to this schema, a mining company with a 'measured' and 'proven' mineral reserve has outlined the exact contours of a mineral deposit and has developed a plan to extract that ore at a profit. Ascendant's estimate was an inferred resource, the entry-point on the CIM framework. Although the Toronto rules are explicit in saying that an 'inferred' estimate bears no relation to economic ore, the classifications imply a direct relationship between time, resource estimates, and geological knowledge. It invokes a temporal horizon of future mineral production. It assumes the progression through time whereby an inferred resource becomes indicated and measured, and a speculated resource becomes an economically viable mineral reserve.

In this regard, there is a symbolic parallel between an inferred resource and a junior mining company. Just as an inferred resource is supposed to mature into a measured one, the term 'junior' implies a process of maturation: the simultaneous becoming of mineral resources and the junior companies attached to them.

According to the CIM framework, Bishimetals' resource estimate, completed prior to the new standards, was a 'historical estimate'. In order to become an 'inferred' estimate, the Toronto rules also required that Ascendant have an outside geologist, a so-called 'Qualified Person', review and sign off on the company's methodology of resource estimation. To do so, Ascendant contracted Micon International, one of the many consulting companies that quickly filled the niche of independent geological auditors after the establishment of the CIM rules in 2000. Micon's audit report, conducted in 2005, found that Ascendant's estimate was 'in compliance with CIM standards' as an 'inferred' resource (Micon International 2005: 30). The classification of an 'inferred' resource is, however, incredibly vague and, in a sense, accommodating. As it contains no guarantee of being upgraded, the real purpose of the audit process is not to judge whether Ascendant's specific techniques of block modelling are legitimate or not, but rather to bestow the classification of an 'inferred' estimate. An inferred estimate itself makes no assurance, and an audit can only confirm that. Indeed, the Micon report acknowledges, in the spirit of disclosure, that all data and information came from Ascendant Copper (Micon International 2005: 4).

Apart from the parallel between an inferred estimate and a junior mining company, there are noteworthy similarities between the conceptual space of an inferred classification and Richard Schechner's descriptions of the creative space of the theatre workshop. According to Schechner, the 'is' of a final performance is preceded by the 'as if' setting of the workshop. This 'as if' mood of the workshop is, Schechner argues, 'something tentative, subjunctive', the space of: 'Let's try that', 'This could work', and 'What would happen if?' (Schechner 1985: 103). A literal sense of 'what if' certainly characterizes what we see of Ascendant Copper's techniques of appraisal: what if we used a larger block size; what if we used a variable radius to define those blocks; and what if we assumed the resource was more abundant at depth, beyond the scope of 600 metres? Even these 'what if' assumptions, made visible in the Micon audit report, only hint at the invisible work of 'what if' that is concealed within Ascendant's claim of proprietary software. Precisely because it makes no guarantees, the very classification of 'inferred resource' permits, and very well encourages, this sort of subjunctive analysis. Indeed, subjunctive – 'relating to or denoting a mood of verbs expressing what is imagined or wished or possible' (Oxford American Dictionary 2010) – reflects the underlying mood of an inferred estimate. Furthermore, in the same way that most of what happens in the workshop is thrown out, the same is true of how the CIM standards define an inferred estimate: '[I]t cannot be assumed that all or any part of an Inferred Mineral Resource will be upgraded to an Indicated or Measured Mineral Resource' (quoted in Micon International 2005: 29).

Despite the workshop-like malleability of Ascendant's 'what if' strategies of resource estimation, the endeavour itself is based upon an 'is' relationship between extracted core material and its subsoil referent. Ascendant's estimate was predicated on the

Table 3. Analysis of copper in the sample boreholes

Hole number	Bishimetals (1990, %)	Ascendant (2007, %)
30	1.77	0.88
29	1.25	0.50
28	0.57	0.33
27	0.46	0.28

legitimacy of Bishimetals' exploration programme. Ascendant assumed that Bishimetals was good at extracting core, but somehow flawed in its secondary analysis of plotting that material into the outlines of a subsoil copper deposit. In this regard, the audit from Micon unearthed noteworthy aspects of the Bishimetals exploration programme. The Micon report notes that there is 'little information' about Bishimetals' 'sample protocol' as well as a lack of 'assay certificates or drill hole logs' (Micon International 2005: 23). More importantly, the audit observes that, 'The laboratory that carried out the testing is not known' (Micon International 2005: 23). Despite these acknowledged uncertainties, the Micon report did affirm that the core data – at least the half that was stored in Quito – 'was still in good condition'. Indeed, according to Micon, 'the core and core boxes have been labeled and maintained in a manner consistent with good industry practice' (Micon International 2005: 22). Even as the audit raised questions about the chemical analysis of extracted rock, it affirmed the material stability upon which the indexical relationship between core and deposit relied. Nevertheless, given the unknowns of Bishimetals' sample protocol and geochemical analysis, Micon suggested that Ascendant re-examine, or re-assay in geological terms, the core material that had been stored by the Ecuadorian government since the mid-1990s.

Given that Ascendant's resource estimate did not fault the technical sampling of Bishimetals' data, but rather how it was interpreted, the expectation of the re-assay was to find the same mineral content as that derived by Bishimetals (Ascendant Copper 2007a). Doing so would not only validate Bishimetals' exploration programme, but also reaffirm the indexical relationship between core and subsoil deposit. In 2007, following Micon's recommendations, Ascendant sampled the four drill holes that went to 600 metres – the four drill holes that played an outsized importance in its own inflated estimate. The copper content found in this analysis, however, was markedly different than that detected in the 1990s (Ascendant Copper 2007a). In fact, it was less than half, on average (Table 3). In response, Ascendant was forced to admit that 'the inferred resource estimate contained in the [2005 audit] of 982 million tons grading 0.89% copper may not be fully representative of the mineralization at Junin' (Ascendant Copper 2007a).

It is unclear what the lower figures would have done to Ascendant's resource estimate. Ascendant, unsurprisingly, did not produce a new estimate to incorporate those numbers. Instead, the company spun a different story about the reliability of the core material stored in a Quito warehouse. Whereas Ascendant had readily promoted the underlying material truth of the core data in 2005, in response to the 2007 assay results, it said the re-examination 'was based on historical data, some of which is now considered to be unverifiable due to deterioration of the core samples over time' (Ascendant Copper 2007a). Avoiding thorny questions about the reliability of Bishimetals' exploration programme, Ascendant instead blamed 'the ravages of time' (Ascendant Copper 2007a). In subsequent reporting, Ascendant also referenced 'the loss or theft' of historic core, the presence of 'termites and open exposure to the elements', and 'the possible mislabeling of core boxes, among other factors' (Ascendant Copper 2007b). With these factors highlighted, Ascendant maintained the 982 million ton assessment. Nevertheless, it was forced to date or link that resource estimate to the 2005 auditing report by Micon International. In the wake of the check-assay programme, Ascendant had to argue repeatedly that its 2005 estimate 'remains current and indicative', an awkward and anachronistic back-dating that is incongruous with the arrows of temporal progress on the CIM framework (Fig. 1).

The juxtaposition of Bishimetals' original core material – whether atrophied or intact – reminds us that the size and content of Junin's copper is not imagined in relation to an underground copper resource. Rather, it is mediated by an assemblage of rock material – half of which, in 2007 anyway, was sitting in wooden boxes in a warehouse in southern Quito. The estimates of Bishimetals and Ascendant are talking *directly* about rocks in boxes, and only *indirectly* about the subsoil rock beneath Junin. To extend the metaphor of performance, Ascendant's practice of resource estimation, the making of a spectacle through the use of block modelling and proprietary software, depended upon the non-performance or the invisibility of the indexical relationship between core and deposit. This assumed equivalence between core and deposit is itself the stage that makes possible Ascendant's dramatic show of potential copper abundance.

Through the spectacle of abundance, Ascendant was able to attract millions on the Toronto Stock Exchange. From the sale of company stock between 2005 and 2007, it generated nearly $26 million in cash – a prime example of how junior companies 'mine the stock market', in the words of industry insider Mickey Fulp (2008). The value of the company's stock, nevertheless, declined significantly during that period. Whereas Ascendant's prospectus in 2005 offered a share price of $1.60 (Ascendant Copper 2005b), in 2007, after the re-assay results, the company offered shares valued at $0.36 (Ascendant Copper 2007b). By March 2008, shares in Ascendant averaged less than 25 cents. By September 2008, the company's stock valued about 2 cents and Ascendant, renamed Copper Mesa, was forced to take out a bridge loan of $2.3 million (Copper Mesa 2009).

As the company's stage of subsoil copper was, literally, degrading, it was through a language of degradation and poverty that Ascendant portrayed the Intag surface landscape and its own role in shaping the region's future.

Prognosis as diagnosis

On 7 July 2004, I watched as Patricio Polanco, one of Ascendant Copper's Ecuadorian representatives, took centre-stage in a small schoolhouse in the Intag parish town of Peñaherrera. In his hand, Polanco held a four-inch section of rock core. I was in Intag that summer conducting research for my Master's thesis. Polanco was joined by two colleagues, also Ecuadorian. This was Ascendant's first public presentation in Intag. They came with a PowerPoint presentation in hand and projected information about Ascendant and its vision of Intag onto a white sheet duck-taped to the wall behind the stage. Polanco faced at least two hundred people from the region. Some sat in the front on white plastic chairs; others on green wooden benches right behind them. Most stood, myself included. Those who could not fit into the room listened through open and broken windows. Children ran around in the school's courtyard. After some general, and brief, remarks about modern mining, Polanco talked about the difference between exploration and production. As the PowerPoint projected the question: 'What phase are we in now?' Polanco answered 'exploration', and explained how this phase lasted fifteen years on average, at least for copper mines. Polanco then held up the small section of rock core. 'This is the only thing we want to extract: samples'. Downplaying the economic significance of the core material, Polanco said that if it were made entirely of copper, it might only be worth two or three dollars. In other words, Polanco emphasized, 'this is not copper that we want to sell', but only 'information'.

Polanco then showed an image of an exploration platform and drill. 'As you all can see', he told the crowd, 'it takes up no more space than a volleyball court – this is all the space we need: nothing more ... we are not doing anything that might produce contamination'. Once exploration was complete, Polanco said, there would be a period of 'detailed study' to determine if Junin was a viable deposit. 'If a mine is not economically feasible, then everything stops', he declared, almost wishfully. As the project in Junin was 'preliminary', Polanco concluded, Ascendant was 'still very far from this phase' of assessing economic viability.

In his presentation with the core samples, Polanco argued that it was premature to worry about the potential impacts of copper mining. This was one of Ascendant's recurring arguments. It featured prominently in a back and forth on an on-line forum sponsored by the Business and Human Rights Centre (*business-humanrights.org*) between Ascendant CEO Gary Davis and Carlos Zorilla, the director of an Intag-based environmental group called Defense and Ecological Conservation of Intag (Decoin). In his letters, Davis continually put off mining as beyond the horizon of debate. 'Of critical importance', Davis emphasized in one letter, 'is the fact that Ascendant does not even have mining properties at this time; it has only *exploration properties*' (August 2006, original emphasis). In response to Decoin's complaint about Ascendant's incomplete environmental impact study (EIS), Davis reiterated that Ascendant 'was only raising money for exploration'. 'As such and until such time as we know whether there even exists an economically developable deposit', Davis continued, 'it is premature to spend significant dollars on a full-blown EIS' (June 2005). This language of 'as such and until such time' underscores Ascendant's attempt to foster what we might call a temporal politics of 'in the meantime' – a time period during which Ascendant could explore the Junin deposit without being questioned about what that exploration would lead to.

In portraying mining as a relevant topic only once exploration was complete, Ascendant tried to shift attention to problems of poverty and environmental degradation. According to the company's logic, the 'object of analysis and target of intervention' (Foucault 1990: 26) for Intag should be issues like deforestation, not the distant prospect of mining. Against the impression of poverty-linked degradation, Ascendant portrayed itself as an entity uniquely suited to address these pressing problems. In doing so, it drew directly and indirectly on images, ideas, and practices associated with the field of CSR. In her ethnography, Dinah Rajak focuses less on the targets of CSR, and more on 'its purveyors, and the apparatus through which [CSR] is deployed and dispensed' (2011*a*: 17) This focus, Rajak argues, offers a window into the self-fashioning of corporate agency and the 'performance of corporate virtue' (2011*a*: 2). Like Rajak, I focus here on the constitutive logics through which Ascendant portrayed itself as a 'responsible corporate citizen' that was 'building an infrastructure of community services and environmental health' (AscendantCopper.com n.d.). In particular, I consider how Ascendant utilized a discourse of diagnosis in its attempts to shift public discourse away from mining and onto what it defined as more immediate environmental issues.

Journal of the Royal Anthropological Institute (N.S.), 67-86
© Royal Anthropological Institute 2016

In the Peñaherrera meeting in July 2004, Cesar Villacis, Ascendant's head of community relations, echoed Polanco's temporal framing in his illusion of future co-operation between Ascendant and local communities. 'In the meantime', Villacis went on, Ascendant was in the process of creating a Local Development Plan catered to Intag's short- and medium-term future. The plan, said Villacis, would emerge through 'a complete diagnostic of the actual situations on-the-ground' for each community and the Intag region as a whole. One of Villacis's co-workers, Dr Sanchez, then took over and emphasized the process of diagnosis – a phrase he repeatedly stressed – to describe what he saw as a vicious cycle of poverty, inefficient agriculture, and indiscriminate logging that imperilled Intag's environmental and human health. 'Each day the region gets drier', Dr Sanchez told the crowd, 'it is moving towards desertification because of the bad habits of the colonists'.

A key part of Ascendant's diagnostic rhetoric was photographic imagery. Ascendant infused its promotional material with images of deforested landscapes in Intag. The company website was replete with pictures of environmental loss, from pastures and deforested hillsides to mules carrying planks of wood, all captioned with expository phrases like 'The area has already suffered heavy deforestation' and 'This is their only savings account, a tree'. Ascendant rehearsed these arguments in the second edition of *El Inteño*, a company publication that circulated within the region. In a prominent article headlined 'Satellite images of deforestation in Intag', Ascendant used what it described as 'the prestigious program Google Earth' to show images of agrarian settlement in the parish of Garcia Moreno, the westernmost section of Intag that includes Junin (*El Inteño* 2007). The 'advance of deforestation', the article said, not only imperilled the forests above Junin, but was also a 'serious threat' to the national park that bordered Intag. The article concluded with further illusions of prognostic expertise: 'We are currently undertaking studies of great technical precision about deforestation in Intag and in future editions we will continue to provide information about this important theme' (*El Inteño* 2007).

In articulating imagery and an urgent temporality of 'advancing deforestation' (*El Inteño* 2007), Ascendant aimed to situate cloud forest biodiversity not in terms of the long-term prospect of mining, but rather in relation to the agrarian livelihoods of those who opposed the project. Ascendant buttressed its verdict that local farmers posed the greatest threat to nature by alluding to Intag's cultural history as non-indigenous, *mestizo* terrain.[5] On its webpage in 2005, Ascendant stated, 'There are no indeginous [*sic*] tribes in the area', only 'Mestizo slash & burn farmers', whose 'presence has continued to expand'. Another Ascendant-sponsored website (which I discuss below) described locals in more blunt terms: 'They're not even indigenous! They've only been here 30 years'. Against the image of pristine forests under threat from slash-and-burn agriculture, Ascendant then claimed that 'the Junin project is perhaps the only viable solution in saving the cloud forests and preserving what remains of the biodiversity' (AscendantCopper.com n.d.).

Ascendant's self-proclaimed viability reflects the way the company used imagery of degradation to undermine the regional environmental group Decoin. Carlos Zorilla, a Cuban-born, US expatriate who settled on a farm in the eastern section of Intag in the 1970s, helped create Decoin in the mid-1990s in response to the presence of Bishimetals. In the fall of 2004, a website called Fraudecoin.org appeared on-line. Under the guise that 'a photo is worth a thousand words', Fraudecoin showed images of deforested hillsides in Intag with captions (in both English and Spanish) like 'Massive Deforestation,

Mining was not the cause!' and 'Junin's Forest, Going, Going, GONE!' (Fraudecoin.org n.d.). These images, according to the website, proved that Decoin was not only a fraudulent group of 'pseudo-environmentalists', but also a corrupt organization that received untold millions with 'no accountability of the money received on behalf of local communities'. In Fraudecoin's telling, local communities, like Junin, did not oppose mining on their own accord but were being 'held captive' by Decoin's radical thinking. Though it bore no official imprint of Ascendant Copper, Fraudcoin's narrative mirrored Ascendant's characterization of Intag and Decoin. In his back and forth with Zorilla on the forum sponsored by the Business and Human Rights Centre, Ascendant CEO Gary Davis continually described Decoin as the primary instigator of opposition to Ascendant. Portraying Ascendant as a company 'committed to doing what is right by the people of Intag', Davis could not understand what exactly Decoin was opposed to. 'It is incredible to us', Davis wrote, 'that Decoin does not support and join us in our efforts to enhance the living standards of those residing in the communities [near the concession]'. In Davis's framework, the company's efforts were not tied to mining, but rather derived from the company itself and the political space of 'in the meantime'.

Against the conjured stage of degraded landscapes and ineffectual institutions, Ascendant justified its presence in Intag and promoted its assorted actions on the ground. In early 2005, the company purchased a large farm in the centre of the valley below Junin. It described the property as a 'model demonstration farm', the centrepiece of the company's programme to teach locals the tenets of sustainable agriculture while instilling values of leadership and conservation. While I was in Junin during the summer and fall of 2005, activity at the farm varied. At one stage, I witnessed between five and ten workers, mostly pro-mining individuals from surrounding communities, building a scenic trail along a river. I also observed workers begin the construction of a greenhouse, vegetable garden, poultry coop, and tree nursery. Towards the end of the year, Ascendant also began to use the farmhouse to host various leadership workshops and occasional visits from company-sponsored medical teams.

According to Ascendant's promotional material, the projects at the model farm were part of an array of programmes that the company spearheaded throughout the area. The tree nursery at the model farm was, Ascendant claimed, one of at least five tree nurseries in different communities that contained not only native species, but also marketable fruit trees (Ascendantcopper.com n.d.). The small chicken coop was, according to Ascendant, part of a project in which eighty-five families were involved (*El Inteño* 2007). The presence of health workers was, the company asserted, part of a programme that, through 2006, treated over 1,500 patients at various health clinics around the parish (*El Inteño* 2007). The empowerment workshops were portrayed by Ascendant as part of a broad commitment to education in which the company's personnel not only lectured to high school students on agronomy and environmental management, but also provided basic necessities like computers and transportation. In addition to these, Ascendant advertised a diverse array of initiatives that included a biogas project, the sponsorship of a soccer league, a waste management scheme for the parish town of Garcia Moreno, and an ongoing infrastructure initiative whereby the company claimed to be 'literally building a sustainable future for the community' (AscendantCopper.com n.d.).

Ascendant promoted these initiatives through a general frame of community engagement. Much of this work, however, was centred in the principal parish settlement of Garcia Moreno and surrounding communities – places further afield from the

Journal of the Royal Anthropological Institute (N.S.), 67-86
© Royal Anthropological Institute 2016

mineral concession, but where support for mining was high to begin with. The model farm below Junin underscores the patchwork geography of Ascendant's on-the-ground presence. Though the farm was literally just below the settlement of Junin, and visible from the local roadblock, Ascendant's Ecuadorian personnel moved judiciously between the farm and the homes of pro-mining households nearby. Through a network of walkie-talkies, the farmers opposed to mining carefully monitored the activities of Ascendant representatives when they were in or near the farm. When company personnel approached the community, locals would assemble to stop those movements. When I arrived in Junin in June 2006, for example, I was told of an encounter that happened two months prior. That May, a group from Junin assembled just below the community to stop a small Ascendant team that had walked up from the demonstration farm and began taking water samples from the banks of the river. According to the company, the water sampling was part of a programme of Community Environmental Monitoring. When confronted, the Ascendant team retreated back to the demonstration farm, but not before two women from Junin tried to the throw the group's leader into the river.[6]

As this confrontation reveals, those opposed to mining were never convinced by Ascendant's discursive separation between exploration and exploitation. Nor did they see the model farm, or the varied projects it symbolized, as anything but instrumental actions geared towards the establishment of an open-pit mine.

Rajak describes CSR as a 'coherent framework of practice and knowledge, which produces and sustains [its] truly discursive power ... in the Foucauldian sense' (2011a: 14). One can appreciate the Foucauldian logic of Ascendant's politics of above-ground prognosis, the way its diagnostic imaging of deforestation informed the company's self-portrayal as an agent of conservation and development. Ascendant's iteration of social responsibility, nevertheless, lacked both the coherence and discursive power Rajak alludes to. Much of the company's articulation of CSR was impulsive, misplaced, and rather amateurish.[7] This is particularly true of the company's environmental discourse. For example, Ascendant's own webpage, for a time, described the company's core values in the following way: 'To hold the human factor in the project as the most important issue. Maximize the conservation of the environment, through achieving sustainable goals. To find the harmony between *conservatism* and profitable production' (AscendantCopper.com n.d., emphasis added). Given the notion that corporate care is premised upon professional expertise, or at least a mastery of environmental discourse, the odd slippage displayed above between 'conservation' and 'conservatism' is noteworthy. Far from establishing a discursive middle ground in which the company could partner with a more mainstream environmental organization, Ascendant initially partnered with The Latin American Foundation for Humanist and Scientific Development. This was an obvious Ascendant creation, as evidenced if not by the awkward name, then by the fact that the foundation and Ascendant had the same office address. Outside of the initial meeting and the circulation of an error-filled planning document that bore the foundation's name, Ascendant never mentioned the foundation again.

In this regard, Ascendant's array of CSR projects registers less as a polished 'performance of corporate virtue' and more as a fantasy world indicative of the theatre workshop. In the same way that Schechner describes the theatre workshop as the space where actors try out and play with different characters and combinations of behaviour, Ascendant's practice of CSR was one of impulsive movements from one project and strategy to another. In other words, if the company's CSR was a theatre performance, the stage would have been overwhelmed by a tangled mess of characters, scripts, and set designs. And like the liminal space of the theatre workshop, much of Ascendant's presence on the ground was equally laden with contradiction. Through its CSR projects, it ended up employing at least two hundred people in the parish of Garcia Moreno (Ascendant Copper 2007b: 11). Even as the company portrayed these projects as geared towards sustainable agriculture, it required its employees to wear stiff yellow boots and blue helmets – gear that symbolized mining labour. When I was in Junin during the winter and summer of 2007, I spent time at the house of one of the principal local anti-mining leaders. His home was in the valley below Junin, on the outside of Ascendant's demonstration farm. While there I would observe the daily transportation of these local workers, festooned in mining gear, as they travelled from the Junin valley towards other properties that Ascendant had purchased further afield. This, it seemed to me, embodied much of what the company was doing: having local employees dress up like miners, but transporting them away from the concession to work on agricultural projects.

Ascendant also paired its projects of local development with the constant threat of violence. Its model demonstration farm reflected this duality. As I observed when I spent time at the home of the anti-mining leader, the presence of armed guards nearly always shadowed the movement of the company's employees and local workers. At some stages these security personnel were attired as such. At other times, as I witnessed in the summer of 2006, the threat of violence came from two unknown men who wore dark helmets and travelled to and from the demonstration farm on a daily basis. A month before I arrived that summer, one of these men, who rode a four-wheeler, threatened the anti-mining leader on the road with his gun. As Welker (2009) shows in her analysis of the practices of CSR in Indonesia, community empowerment often exists alongside threats and acts of violence. Ascendant's mode of community engagement reiterates this point. Thus, while the company engaged its supporters through a practice of empowerment and development, its stance towards local opposition was always shadowed by the presence of armed guards.

Nevertheless, Ascendant even characterized its more overt displays of force through a lens of conservation and development. In December 2006, after more than two years of actively promoted community development and empowerment had yielded no material progress, an Ascendant-sponsored paramilitary squad, comprised of ex-police and military, squared off with local opposition at the roadblock. According to Ascendant, the squad was really a group of agricultural experts that aimed to 'create an ecological conservation area' and to 'develop sustainable activities as part of an overall community development plan' (Ascendant Copper 2006b). The episode ended badly for the company. Locals not only refused to budge from the roadblock, but they actually detained the paramilitary group two days later when they became lost in the woods above Junin. The squad, almost sixty in total, readily gave up their guns and walked themselves into the community church, where they remained captive for the next week. Six months after the paramilitary fiasco, Ascendant began to reduce its presence in the region and was completely gone by the end of 2008.

From the perspectives of investors, capital invested in Ascendant was premised upon the future extraction of copper. It was not money meant to sit still, but rather capital given to Ascendant for them to do something. The survival of junior mining companies depends on momentum; it depends on their ability to promote a narrative of linear progress. This linear assumption is evident in the arrows of the CIM framework on mineral classification (Fig. 1). Through the increase of geological knowledge over time, 'inferred' mineral resources are expected to become 'probable' and 'proven'. As I have described above, Ascendant employed a diverse set of strategies to access the mineral concession. Apart from positing exploration as benign information-gathering, it promoted an image of Intag as degraded social and environmental terrain, against which the company depicted itself as an institution of conservation and development. Ascendant's rhetoric and practice of CSR, from this perspective, reflect an underlying strategy to undermine local opposition and enlist local support.

On the other hand, I also argue that CSR was useful for Ascendant in a more instrumental fashion. It gave the company something to do. CSR, in this sense, became an outlet for Ascendant to spend investment capital. Outside of salaries and administrative costs, it categorized all of the money that it spent in Intag as 'exploration'. Exploration, in the geological sense, represents linear movement, and that is precisely what Ascendant's images of CSR portrayed. Be it an image of a doctor checking on a patient, an Ascendant employee unveiling a new computer to high school students, an Ascendant representative standing in a tree nursery, or an Ascendant engineer training local environmental monitors, Ascendant's images of corporate improvement all showed a company that was doing things. These were 'projects in action' (AscendantCopper.com n.d.) that not only showed progress, but also suggested an image of corporate stability and corporate agency. Indeed, as the company promoted on its website through 2007, 'Without unyielding corporate social responsibility and environmentally sustainable practices Ascendant would not be the successfully growing company that it is today'. Of course, by the end of 2007, Ascendant was not a growing corporation, but a stagnating one faced with deteriorating stock prices, deteriorating rock core, and local opposition that refused to move.

Conclusion

The dual meanings of prognosis, as both forecast and diagnosis, provides a useful lens with which to consider the divergent strategies and temporalities that Ascendant Copper employed in its attempts to garner investment and undermine local opposition. Ascendant's articulation of an inferred copper resource was based on a temporal horizon of future mineral extraction. In this sense, an inferred resource has no present. It only exists in relation to an economically viable deposit, and the future prospect of mining. Ascendant's rhetoric of diagnosing and addressing environmental degradation, on the other hand, was a temporal focus on the immediate present. Put another way, we could describe Ascendant's politics of prognosis as one in which the company tried to fill the empty present of an inferred resource with images and practices of conservation and development. These variations of prognosis did not occur in a vacuum. Ascendant promoted a spectacle of copper abundance in relation to the epistemological framework of the Toronto rules of resource classification. Likewise, Ascendant presented the spectacle of sustainable development by drawing on images and ideas of corporate social responsibility. Despite their contradictions, Ascendant's promotion of sustainable development relied on the company's ability to sell the spectacle of subsoil abundance: its output of CSR depended upon financial input that was geared towards subsoil wealth.

In highlighting this connection between subsoil and surface enactments for a company like Ascendant Copper, I have furthered recent discussion on the conditions of possibility of both CSR and corporations more broadly. The tensions and connections between Ascendant's rhetoric of subsoil and surface prognosis are not unique to Ascendant Copper, or even junior mining companies. They are a constitutive feature of corporations whose economic futures depend upon finite mineral resources. Indeed, as Rajak acknowledges in the introduction to her book (2011a: 15), sharp declines in the commodity price of platinum led to severe reductions in Anglo-Americans' CSR commitments. Thus, while scholars approach CSR through the tensions between the pursuits of profit and social/environmental betterment, in our analysis of the extractive industries in particular, we should not lose sight of how that tension is itself mediated by the subsoil and its commodity outputs. Indeed, I would argue that is precisely through an analysis of how subsoil resources come into being that we can better appreciate Welker's argument to approach 'corporations as inherently unstable and indeterminate, multiply authored, always in flux, and comprising both material and immaterial parts' (2014: 4).

In this regard, even as theories of performance inform literatures on resource materiality and corporate behaviour, a more apt framework may be the theatre workshop. In relation to subsoil copper, the workshop underscores the epistemological licence that goes into the making of an 'inferred' copper resource. In relation to Ascendant's stance towards Intag's surface landscape, the workshop allows analysis of the company's CSR performance in a way that highlights the messiness and contradiction therein. The workshop allows us to think about the Foucauldian logic without also assuming a Foucaldian effect. In analysing both forms of Ascendant's prognosis, prediction and diagnosis, subsoil and surface, we see less of a sophisticated corporate machine and more a junior resource company awkwardly trying to make its way on stage for the opening night.

NOTES

This article builds on research that was funded by the National Science Foundation Program in Science and Technology Studies, the Yale MacMillan Center Dissertation Research Grant, and the Yale Tropical Resources Institute. I thank Michael Dove, Carol Carpenter, Jessica Barnes, Jane Guyer, Karen Hebert, Douglas Rogers, Elizabeth Ferry, K. Sivaramakrishnan, and the anonymous reviewers for their insightful comments on drafts of this paper. I am grateful to Glen Kuecker, Carlos Zorilla, Mary Ellen Fieweger, and the communities in Intag for making my research in the region possible over the past decade.

[1] Within these edited volumes, see the excellent ethnographies of mining conflicts in southern Ecuador by Moore and Velasquez (chapter in both books) and by Warnaars (chapter in Bebbington & Bury 2013). Davidov (2014) provides an insightful account of some of the dynamics of the conflict in Intag.

[2] Almost all of my research in Intag has been based in the community of Junin and the surrounding valley, research that has been aligned with local opposition. (On the positionality of research within mining conflicts, see Kirsch 2002 and Kneas 2005.) Part of this work (August-November 2005, July 2006, January 2007) was in conjunction with an international observers programme that I have been involved with, one spearheaded by a former undergraduate professor (Kuecker 2007). I conducted pre-dissertation research during June-July 2007, and June-July 2008. This was followed by more than a year of ethnographic research that began in August 2009 and which situates local divisions over mining in relation to shifting patterns of agrarian settlement and associated dimensions of class and power (Kneas 2014). I first travelled to Intag in 1999 as a semester abroad student and managed a cloud forest reserve in the region between 2000 and 2004, a time during which I also worked closely with regional environmental organizations. I spent the summer of 2004 in Intag observing (and filming) Ascendant's initial public presentations as part of my Master's thesis project (Kneas 2005).

[3] In July 2008, Ascendant changed its name to Copper Mesa Mining Corporation.

[4] This comes from a forum on Canadian mining companies on investorshub.com, and a thread dedicated to Ascendant Copper. Deep_Digger's comment came on 23 March 2007: *http://investorshub.advfn. com/Ascendant-Copper-TSX-acx-8339/*.

[5] The mining conflict in Intag has not had distinct articulations of indigenous identity (Li 2000), nor is it one with historical undercurrents of indigenous-*mestizo* land conflicts like other regions of the Ecuadorian highlands (Clark & Becker 2007). Although it was home to native groups in the precolonial period (Salomon

1997), like other frontier areas in western Ecuador (Clark 1998*a*; 1998*b*) and in Colombia (Applebaum 2003), Intag emerged as a distinct region within mid-twentieth-century Ecuadorian national space through metrics of race, territory, and agrarian settlement that culturally marked Intag as *mestizo* terrain. Apart from a few self-identifying indigenous communities in older sections of Intag, the majority of the population in Intag identifies as *mestizo*. In the parish of Garcia Moreno, less than 3 per cent of the population self-identified as indigenous in the 2010 census. Within the communities that are closest to the mining concession, and that are most opposed to mining, many farmers are first- and second-generation Colombian.

[6] Part of this anger came from the fact that this same woman, a year prior, had spent the night in the Junin tourist cabin claiming to be a tourist from Quito.

[7] Bebbington (2012*a*: 81) makes a similar observation about the haphazard response of a junior mining company in Peru.

REFERENCES

APPEL, H. 2012. Offshore work: oil, modularity, and the how of capitalism in Equatorial Guinea. *American Ethnologist* **39**, 692-709.

APPLEBAUM, N. 2003. *Muddied waters: race, region, and local history in Colombia, 1846–1948*. Durham, N.C.: Duke University Press.

ASCENDANT COPPER 2005*a*. Preliminary prospectus: initial public offering (23 June).

———— 2005*b*. Prospectus (14 October).

———— 2006*a*. Ascendant reports on re-logging of core samples at Junin Copper Porphyry Project, Ecuador. Press release (9 March).

———— 2006*b*. Ascendant Copper Corporation responds to NGO allegations regarding its Junin Project. Press release (20 December).

———— 2007*a*. Results of check assaying program on Junin property (Ecuador). Press release (11 June).

———— 2007*b*. Short form prospectus (14 June).

ASCENDANTCOPPER.COM n.d. Website defunct, archive of webpage material in possession of author. Last accessed December 2009.

BAKKER, K. & G. BRIDGE 2006. Material worlds? Resource geographies and the 'matter of nature'. *Progress in Human Geography* **301**, 5-27.

BARAD, K. 2003. Posthumanist performativity: toward an understanding of how matter comes to matter. *Signs* **28**, 801-31.

BEBBINGTON, A. 2012*a*. Social conflict and emergent institutions: hypotheses from Piura, Peru. In *Social conflict, economic development and extractive industry: evidence from South America* (ed.) A. Bebbington, 65-88. London: Routledge.

———— (ed.) 2012*b*. *Social conflict, economic development and extractive industry: evidence from South America*. London: Routledge.

———— & J. BURY (eds) 2013. *Subterranean struggles: new dynamics of mining, oil and gas in Latin America*. Austin: University of Texas Press.

BENSON, P. 2008. Good clean tobacco: Philip Morris, biocapitalism, and the social course of stigma in North Carolina. *American Ethnologist* **35**, 357-79.

———— & S. KIRSCH 2010. Corporate oxymorons. *Dialectical Anthropology* **34**, 45-8.

BILLO, E. 2015. Sovereignty and subterranean resources: an institutional ethnography of Repsol's corporate social responsibility programs in Ecuador. *Geoforum* **59**, 268-77.

BRIDGE, G. 2009. Material worlds: natural resources, resource geography and the material economy. *Geography Compass* **3**, 1217-44.

BURY, J. & A. BEBBINGTON 2013. New geographies of extractive industries in Latin America. In *Subterranean struggles: new dynamics of mining, oil and gas in Latin America* (eds) A. Bebbington & J. Bury, 27-66. Austin: University of Texas Press.

BUSCHER, M., D. GOODWIN & J. MESMAN 2010. Ethnographies of diagnostic work: introduction. In *Ethnographies of diagnostic work: dimensions of transformative practice* (eds) M. Buscher, D. Goodwin & J. Mesman, 1-14 . New York: Palgrave Macmillan.

CLARK, A.K. 1998*a*. Race, 'culture', and mestizaje: the statistical construction of the Ecuadorian nation, 1930–1950. *Journal of Historical Sociology* **11**, 185-211.

———— 1998*b*. Racial ideologies and the quest for national development: debating the agrarian problem in Ecuador (1930–50). *Journal of Latin American Studies* **30**, 373-93

———— & M. BECKER (eds) 2007. *Highland Indians and the state in modern Ecuador*. Pittsburgh: University Press.

COPPER MESA 2009. Management's discussion and analysis (September).

DAVIDOV, V. 2014. Land, copper, flora: dominant materialities and the making of Ecuadorian resource environments. *Anthropological Quarterly* **87**, 31-58.

DGGM 1985. Junin miner prospect: geological, geochemical, and geophysical investigation. Center for Geological-Mining-Environmental Information (CIGMA). Ministry of Non-Renewable Resources (Quito, Ecuador). Call number: GMMPI 1612.

EL INTEÑO 2007. Imagenes satelitales sobre la deforestacion en la zona de Intag. **2** (April), 8.

FERRY, E. & M. LIMBERT 2008. *Timely assets: the politics of resources and their temporalities.* Santa Fe, N.M.: School for Advanced Research Press.

FOUCAULT, M. 1990. *The history of sexuality: an introduction* (trans. R. Hurley). New York: Vintage.

FRAUDECOIN.ORG n.d. Website defunct, archive of webpage material in possession of author. Last accessed December 2006.

FULP, M. 2008. Mining the stock market. MercernaryGeologist.com, 20 October (available on-line: *http://www.goldgeologist.com/mercenary_musings/musing-081020-Mining-the-Stock-Market.pdf*, accessed 7 January 2016).

——— 2009. Exploration in emerging environments. MercenaryGeologist.com, 18 May (available on-line: *http://www.goldgeologist.com/mercenary_musings/musing-090518-Exploration-in-Emerging-Environments.pdf*, accessed 7 January 2016).

GUYER, J. 2010. The eruption of tradition? On ordinality and calculation. *Anthropological Theory* **10**, 123-31.

INGOLD, T. 2012. Towards an ecology of materials. *Annual Review of Anthropology* **41**, 427-42.

JENKINS, H. 2004. Corporate social responsibility and the mining industry: conflicts and constructs. *Corporate Social Responsibility and Environmental Management* **11**, 23-34.

KIRSCH, S. 2002. Anthropology and advocacy: a case study of the campaign against the Ok Tedi Mine. *Critique of Anthropology* **22**, 175-200.

——— 2014. *Mining capitalism: the relationship between corporations and their critics.* Berkeley: University of California Press.

KNEAS, D. 2005. Contesting copper: documentary film, research, and mining in Ecuador's Intag region. *Tropical Resources: The Bulletin of the Yale Tropical Resources Institute* **24**, 15-20.

——— 2014. Substance and sedimentation: a historical ethnography of landscape in the Ecuadorian Andes. Ph.D. dissertation, Yale University.

KUECKER, G. 2007. Fighting for the forests: grassroots resistance to mining in northern Ecuador. *Latin American Perspectives* **153**: 34, 94-107.

LI, T. 2000. Articulating indigenous identity in Indonesia: resource politics and the tribal slot. *Comparative Studies of Society and History* **42**, 149-79.

LOWELL, D. 2001. Junior exploration companies: their role in future exploration. *Engineering and Mining Journal* (November) (available on-line: *http://connection.ebscohost.com/c/articles/5576292/junior-exploration-companies-their-role-future-mineral-exploration*, accessed 17 January 2016).

METAL MINING AGENCY OF JAPAN 1998. Informe final sobre la exploracion mineral en la area de Imbaoeste, Republica del Ecuador. Center for Geological-Mining-Environmental Information (CIGMA). Ministry of Non-Renewable Resources (Quito, Ecuador). Call number: GMMPI 1609.

MICON INTERNATIONAL 2005. Ascendant Copper Corporation, technical report: review of the quartz porphyry-hosted copper-molybdenum mineralisation at Junin, Otavalo, Ecuador.

PIEDRA, J. & L. ARRATA 1987. Report on the investigation of the geology and ore deposit at the Junin area in Imbabura province of Ecuador. Center for Geological-Mining-Environmental Information (CIGMA). Ministry of Non-Renewable Resources (Quito, Ecuador). Call number: GMMPI 1611.

POWER, M. 1997. *The audit society: rituals of verification.* Oxford: University Press.

RAJAK, D. 2010. 'HIV/AIDS is our business': the moral economy of treatment in a transnational mining company. *Journal of the Royal Anthropological Institute* (N.S.) **16**, 551-71.

——— 2011*a*. *In good company: an anatomy of corporate social responsibility.* Stanford: University Press.

——— 2011*b*. Theatres of virtue: collaboration, consensus, and the social life of corporate social responsibility. *Focaal* **60**, 9-20.

RICHARDSON, T. & G. WESZKALNYS 2014. Resource materialities. *Anthropological Quarterly* **87**, 5-30.

ROGERS, D. 2012. The materiality of the corporation: oil, gas, and corporate social technologies in the remaking of a Russian region. *American Ethnologist* **39**, 284-96.

ROLSTON, J.S. 2013. The politics of pits and the materiality of mine labor: making natural resources in the American West. *American Anthropologist* **115**, 582-94.

SALOMON, F. 1997. *Los yumbos, niguas, y tsatchila o 'colorados' durante la colonia espanol.* Quito: Abya Yala.

SCHECHNER, R. 1985. *Between theater and anthropology*. Philadelphia: University of Pennsylvania Press.

SCHIEFFELIN, E.L. 1998. Problematizing performance. In *Ritual, performance, media* (ed.) F. Hughes-Freeland, 194-207. London: Routledge.

STRATHERN, M. (ed.) 2000. *Audit cultures: anthropological studies in accountability, ethics and the academy*. New York: Routledge.

THIEME, T., H. BEDI & B. VIRA 2015. Geographies of corporate practice in development: contested capitalism and encounters. *Geoforum* **59**, 215-18.

TREVETHAN, N. 2005. Mining exploration spending trebles since 2002. *Reuters* (10 November).

TSING, A. 2000. Inside the economy of appearances. *Public Culture* **12**, 115-44.

WELKER, M. 2009. 'Corporate security begins in the community': mining, the corporate social responsibility industry, and environmental advocacy in Indonesia. *Cultural Anthropology* **24**, 142-79.

——— 2014. *Enacting the corporation: an American mining firm in post-authoritarian Indonesia*. Berkeley: University of California Press.

———, D. PARTRIDGE & R. HARDIN 2011. Corporate lives: new perspectives on the social life of the corporate form. *Current Anthropology* **52**: S3, S3-S16.

Abondance en sous-sol et absence en surface : une jeune société minière et ses performances de pronostic dans le nord-ouest de l'Équateur

Résumé

Depuis longtemps, les investissements miniers mondiaux sont marqués par le spectacle et les performances. L'auteur examine ici les stratégies performatives d'une jeune société minière nommée Ascendant Copper au cours de sa prospection de cuivre dans le nord-ouest de l'Équateur au milieu des années 2000. Il se concentre sur les efforts contradictoires d'Ascendant pour attirer les investissements miniers. Elle déploie une imagerie d'abondance du sous-sol, tout en minimisant l'échelle et les répercussions de l'exploitation minière face à l'opposition locale. L'article aborde les stratégies divergentes de la société comme le reflet de la double signification du pronostic, qui est à la fois prévision et diagnostic. Il commence par examiner par quels moyens Ascendant a mis en avant la richesse en cuivre du sous-sol dans le cadre de la classification des ressources établie par la Bourse de Toronto. Il analyse ensuite la manière dont la société a utilisé le discours de la responsabilité sociale d'entreprise pour dépeindre sa présence locale en termes de conservation et de développement. En analysant cette jeune entreprise à la fois par rapport au sous-sol et à la surface, l'auteur met en lumière la matérialité de la responsabilité sociale d'entreprise tout en soulignant la situation précaire des jeunes entreprises, qui apparaît rarement dans les descriptions anthropologiques des entreprises.

5

Mines and signs: resource and political futures in Bangladesh

Nusrat Sabina Chowdhury *Amherst College*

In Bangladesh in 2007-8, speculation about an energy catastrophe thrived alongside utopic visions of democracy. Two events consolidated the despair and hope for collective national futures: a movement against possible coal mines in Phulbari in the northwest and a nationwide political emergency. They are the focus of this essay. By bringing resource crisis and political crisis together, I argue, at one level, that resource and democracy have become two exemplary sites of thinking towards the future, and, at another, that the gaps in their varied imaginations are potential sites of politics. Drawing on my ethnography of the anti-mining protests in Phulbari, I analyse the relationship between signification and anticipation in Bangladeshi politics. In so doing, I identify those emergent political moments that challenge the anticipatory politics of the state and the energy company.

In 2007-8, walking around the small township of Phulbari in northwest Bangladesh, one routinely came across anti-mining slogans scattered across mud huts, the walls of *paka* (brick) houses, and on the backs of rickshaws. A violent scramble for resources had unfolded here in 2006 when local sentiments against mining, particularly the forced displacement that this would entail, led to protests against the multinational mining company then called Asia Energy and the Bangladeshi state that represented its interests. One slogan, relatively small and barely legible, caught my eye as I came out of a house in one of the villages more than a year later (Fig. 1). The message, hand-written, smudged, and on the verge of being covered up by newer inscriptions, was clearly lacking the aesthetic conventions of political graffiti. It reiterated a sentiment echoed in relative abundance across Phulbari: 'We don't want coal mines; we don't want coal mines destroying agricultural land' (*koyla khuni chaina/krishijomi dhwangso kore koyla khuni chaina*).

This particular representation of the movement's core demand stood out for me because of its articulation in the local dialect, which never made it to the official script of the protest. *Khoni*, the Bengali word for 'mine', became *khuni*, 'murderer' or 'killer'. The *double entendre*, neatly folded into a mere spelling mistake, rang eerily true when read against the backdrop of the violence that had erupted here since the arrival of

Figure 1. Anti-mining slogan on the wall of a house in Phulbari. (Photograph by the author.)

Asia Energy, including the death of three men, and the looming prospect of mass relocation.[1] The misrecognition of the 'mine' as the 'murderer' is my point of departure for writing about the slippages between other signs of collective futures that struggled for prominence around the topics of resource and democracy in Bangladesh. Drawing on personal narratives collected in Phulbari, where I lived in 2007-8, as well as the texts in circulation by the multinational, the media, and the residents of the mining area, this essay explores the relationship between signification and anticipation in Bangladeshi politics.

In Bangladesh in the mid-2000s, speculation about an impending energy catastrophe thrived alongside renewed hopes for democratic governance. Two events that helped consolidate the despair and hope around resource and politics form the basis of my essay. The first was a massive upheaval against open-pit coal mining in Phulbari, which culminated in violence and (partial) victory for the protesters. Collective grievances reached a climax on 26 August 2006 when three young men were killed in a collision between the paramilitary and the demonstrators. The latter opposed the forcible displacement of more than a hundred thousand people, the method of extraction and the export of coal, and the collusion of the state and a foreign mining company, the UK-based Asia Energy (Bedi 2015; Luthfa 2011; Muhammad 2007). Amid escalating tensions, the company had to suspend its activities in Phulbari. The government agreed to sign a treaty with the local people promising, among other things, monetary compensation for the victims and/or their families.

Phulbari tied together anxieties about an unprecedented energy deficit with uncertainties about the seemingly unmanageable corruption and volatility of national politics. Things took a particularly violent turn when a lack of consensus between the two mainstream political parties in late 2006 regarding the national elections led to fatalities in the streets, and ultimately resulted in a nationwide state of emergency – the second of the two events. Declared in January 2007, the Emergency was to restore order and with it a cleaner and more robust democratic culture. In this essay, I incorporate resource crisis and political crisis in Bangladesh within the same analytical frame. I

Journal of the Royal Anthropological Institute (N.S.), 87-107
© Royal Anthropological Institute 2016

argue that resource and democracy have become two exemplary sites of anticipation, 'of thinking and living towards the future', which has implications for understanding emergent political movements in the Global South (cf. Adams, Murphy & Clarke 2009).

I pay attention to two kinds of resources, broadly defined. One is the mineral object itself: 572 million tons of coal has been found in the Dinajpur region in northwest Bangladesh, the location of the Phulbari Coal Project. The coal sequence, 14 to 45 metres thick and 150-250 metres below the surface, is most profitably extractable, as expert knowledge would have it, by an open-cut mining method.[2] The conflicts over its potential use- and exchange-values fed into the vibrant resistance movement. The other resource is made up of what I call the protest's semiotic culture. By this, I mean the culturally determined and persistently creative ways in which people made, read, and resisted signs, often though not always in written, documentary form. As Webb Keane (2005) suggests, signs entail sociability, struggle, historicity, and contingency. They are situated within a world of consequences rather than as coded messages. In approaching them as another kind of resource, my ethnography supports the claim that material politics is always semiotically mediated, and, more crucially, the slippages between materials and their signs are sites of unpredictable and 'imperceptible' politics (cf. Papadopoulos, Stephenson & Tsianos 2008). This is the politics that arises from the emergence of the miscounted, 'those who have no place within the normalising organisation of the social realm' (Papadopoulos et al. 2008: 67). These are the imperceptible potentials that have been generally misrecognized and translated into the given terms of representation. They are my main focus in this essay.

A rise in prognostic politics characterizes the current global economic arrangement. 'The future' has turned out to be a form of political and technical creativity – a cultural fact, as Arjun Appadurai (2013) would say – open to states, corporations, and ordinary citizens alike (cf. Mathews & Barnes, this volume). It creates material trajectories of life that unfold as anticipated by various speculative processes, what Vincanne Adams, Michelle Murphy, and Adele E. Clarke call 'anticipatory regimes' (2009: 248). These regimes of anticipation are historically specific, culturally conditioned, and transnationally mobile. While conjuring the possibility of new futures is nothing new, anticipation can also become a hegemonic formation – 'big business' in more ways than one – as the large development infrastructures or resource extraction in the Global South reminds us. In thinking of the future, I question what critical possibilities might open up in seeing anticipation as both opportunity and tyranny. My research in Bangladesh urges me to pay attention to people's compulsions to tack between despair and hope, 'to live as subjects in the domain of the not-yet' (Adams et al. 2009: 250). In this I agree with Jamie Cross that 'these dreamed-of futures have lives of their own and material effects upon which our current global moment depends' – a moment he describes as 'an economy of anticipation' (2014: 8). The possibilities enfolded in the idea of anticipation are there in the gap between an ethic of possibility and an ethic of probability of which Arjun Appadurai writes (2013). The ethics of probability are the dimensions of speculation, crisis, and value that inform our thinking on the neoliberal forms of capitalism. The ethics of possibility, in Appadurai's words, are those ways of thinking, feeling, and acting that increase the horizons of hope, expand the field of the imagination, produce greater equity in people's capacity to aspire, and widen the field of informed, creative, and critical citizenship (2013: 295). The scenes from my fieldwork that populate this essay are political openings that surface in the struggle between these two ethical orientations towards the future.

Journal of the Royal Anthropological Institute (N.S.), 87-107
© Royal Anthropological Institute 2016

Ongoing conversations in the anthropology of corporations, the global extractive industry, and contemporary capitalisms guide my observations (Gardner 2012; Gidwani 2008; Kirsch 2014a; Shever 2012; Tsing 2005). This body of work demystifies the language of corporate social responsibility, dismantles the idea of the corporation as a monolith (Kirsch 2014b; Welker 2009), and challenges the approach to capitalism as an internally coherent series of laws. From tracing the interconnections between national and global transformations to attending to the intimate and the affective, ethnographies of environmental, economic, and political futures document connections and contradictions that offer a comparative framework for my understanding of resource and political futures in Bangladesh (Cross 2014; Morris 2008; Shah 2010; Witsoe 2013). Although Asia Energy as a foreign mining company is integral to this story, here I focus instead on imperceptible politics (Papadopoulos *et al.* 2008), whose meanings or effects are never completely stable. This instability produces a space that is carved out of the expansive dominance of the state, the economic power of the corporation, and the passionate rhetoric of the protest movement.

In what follows, I locate some of these political potentialities in Phulbari's daily life. I do so primarily by analysing signs that did not always fit neatly into either the official narrative about Phulbari's wealth or the formalized political discourse that opposed it. A quickly forgotten misspelling, a collective attack on a stranger managed as 'mob' violence, or a counter-reading of the energy company's investors' virtual chats are some of my examples. Analytically, I document the different perceptions of the value of coal in everyday imagination that countered a well-known discourse of scarcity by that of aesthetic abundance. By widening the focus to include national politics, I identify some of the ways in which people in Phulbari challenged the anticipatory politics of the state. For this, I introduce the term 'eyewash' (in its English original), which was popular among many of my interlocutors. I present it as a conceptual other of transparency, the mantra of the Emergency government.

The writing on the wall

Phulbari is a township in Dinajpur district close to the country's northwestern border with India. It is also locatable, figuratively, on a decades-long path towards the privatization of the energy sector (Khan 2006; Muhammad 2014). Large deposits of bituminous coal attracted an open-cut coal-mining project supervised by Asia Energy Corporation, Pty Ltd – a British company.[3] The location of the township correlated neatly with the centre of what was to be the mine's operating area: Phulbari, one of the four *upazilas* (sub-districts) to be affected, was the eye of the storm. Using this coal for power generation would mitigate the Bangladeshi energy deficit for the next fifty years, amounting to nearly half of the national daily energy production. In 2006, homegrown anti-mining agitations that increasingly boasted extra-local and international support gained momentum. When the suppression of protests by the authorities created enough tension, Asia Energy Corporation abandoned its Phulbari office, temporarily shelving the project and leaving the resistance movement in critical abeyance.

Given the densely inhabited and low-lying landscape of the Dinajpur region, the human and ecological costs of the coal project are high (Kalafut & Moody 2008; Muller & Moody 2009). It is also the only mining operation of Asia Energy Corporation (Bangladesh), Pty, Ltd, a fully owned subsidiary of Asia Energy PLC. It was founded in 1997 and renamed as Global Coal Management Resources (GCM) in 2007. It received fiscal support from equity investments of well-known private financial institutions, and at one point sought public sector funds from the Asian Development Bank. With 6 per cent royalty for Bangladesh, between 75 and 80 per cent of the coal was planned for export through the Sundarban mangrove forests in the southwest (FE Investegate 2007; Muhammad 2014).

Phulbari sits on fertile agricultural land. The mining project threatens food security, among other things, by possible desertification of a major rice-growing area. As recently as 2014, a report by the United Kingdom's National Contract Point (NCP) found partial breaches in GCM's obligation to 'develop and apply self-regulatory practices and management systems that foster a relationship of confidence and mutual trust' (Haigh 2015).[4] The grassroots movement opposing the project has been the only success story of similar mobilizations that lay bare the collusion between multinationals and Bangladesh, a country that has been unabashedly loyal to neoliberal policies for decades (Gardner 2012; Muhammad 2015; Sobhan 2005).[5] In 2010, WikiLeaks reported a communication between the US Ambassador to Bangladesh, James Moriarty, and Bangladesh's chief energy adviser in which the former aggressively pushed the government of Bangladesh to realize the project. His interests were not surprising given the 60 per cent of US ownership in it (Karim 2010). The leaked cables confirmed the suspicions that many people in Phulbari shared with me about the kinds of invisible alliances that they imagined to be at work in deciding their fate. It was a story, they believed, where the places and agents in question remained beyond their purview, the signs for which, however, were there for all to see.

The reason I foreground signs when thinking of futures is also in part due to the resource in question. Coal in Phulbari is a subsurface wealth invisible to the naked eye. The knowledge of its existence had not come with any material trace or visual cue for those who happened to spend their lives above this accumulated source of value. Contrast this with Delwar Hussain's description of another Bangladeshi township at the heart of a routine coal trade with India. Hussain says this about the daily life of this raffish border town in the northeast:

> It is teeming with thousands of dark-skinned, gaunt labourers, a fretwork of bones protruding out of their sweaty, taut muscles, their backs and hamstrings tight in desperate efforts. I jump out of their way as they push, lug, heave and pull carts laden with bulky sacks of coal towards the river. Soot has ground itself into their skin and it clouds the paths they walk on ... [E]very so often heavy trucks lumber into Bangladeshi Boropani with coal on their backs (Hussain 2013: 3).

Phulbari, in contrast, is mostly a vast expanse of lush paddy fields dotted by shady villages. As a site of development, it is still being framed in terms of potentiality: a projection of prosperity into the future that tied its coal mines inextricably to the prosperity of the nation. During my research, the state's list of failed services included a huge deficit in power-generating energy. It caused 'load shedding' – to use the term familiar across South Asia (Kale 2014) – for hours on end, and in the summer months plunged the country's major cities and peripheries into routine bouts of darkness, thus calling forth a visual aporia in yet another sense.

The incessant power cuts became proleptic signs of a dark national future. Like population, poverty, and other indices of delayed development, load shedding emerged as a source and a symptom of a malaise that afflicted a nation better known for natural disasters than natural wealth. In a typically blithe comment, a former state minister for energy condensed the two when he stressed the urgency to mine coal at a

Journal of the Royal Anthropological Institute (N.S.), 87-107
© Royal Anthropological Institute 2016

roundtable on the energy crisis that I had attended in the capital in 2008. 'I have been experiencing load shedding since my school life', he exclaimed, 'but now it seems like a real Sidr'. Sidr was a Category 5 tropical cyclone that made landfall in the coastal areas of southern Bangladesh in November 2007 and claimed around ten thousand lives. Those in Phulbari talked about seeing coal shining bright near the boreholes out in the field. They mentioned the gold mines that they had heard lay deep under the earth across the country. Confusing two kinds of subterranean wealth, they appraised coal on the standard of another fetishized mineral: gold. The generous gift of cash, seeds, blankets, television sets, greeting cards, and calendars distributed to the local clubs and the residents of the mining area generated suspicions about the company's motives, while the confrontations between curious or angry villagers and the company guards stationed around the drilling sites often led to violence. Over time, the gifts became telltale signs of the heightened precarity of an uncertain future (Papadopoulos *et al.* 2008).[6]

Phulbari has been a textbook case of transnational connections. Travelling through these circuits of communication, signs mediated hope, despair, risk, and anxiety. They gave credence to people and events that remained hidden from view, acting as they did as promissory notes of a well-lit national future or as moral claims to popular struggle and sacrifice. Take, for instance, the following misreading of signs that led to political action. I am talking about the physical assault of a young man who had travelled to Phulbari from Dhaka, the capital, in 2006 as a leader of a documentary film crew. He was the younger brother of a well-known filmmaker who had sent his team to gather footage of the protests. While there, he was unsuspectingly sporting a T-shirt with the logo of a film festival featuring the word *Asia*. *Asia*, a sign by then familiar owing to the presence of Asia Energy, was enough to anger the crowds, who attacked the man on suspicion of affiliation with the company. Influential activist leaders had to intervene to save him from public wrath. One woman who was part of the 'mob' had this to say:

> *Woman*: We would've killed him. The situation was such that we would've killed him. If each one of us had hit him, he would've been dead.
> *Author*: Did you see the *genji* [T-shirt]?
> *Woman*: Yes! They, like, they looked like they had no life left in them.
> *Author*: What were they doing?
> *Woman*: [The members of the film crew] just came with cameras . . . They knew someone. When they called [that person], he came. Then they checked with [a leader]. [Two other leaders] had to verify if he was someone from [Asia] *energy* or if he was only wearing the *genji*.

At this moment of recognition, *Asia* is an icon; its relationship to the object it stands for is one of resemblance. The word seen on an outsider's clothing and that found on the corporation's logo, such as in the greeting card shown in Figure 2, signifies the same object, a controversial energy company. Icons require guidance in order to determine how exactly they are similar to their objects. This guidance is enmeshed in the dynamics of social value and authority (Keane 2005). The crowd which vandalized the company's office and attacked its employees had located on the body of an urban youth a sign that was not read the same way by all, if 'read' is at all the right verb in this context. It is curious how the woman who confirmed seeing the T-shirt, and most likely did not know the meaning of either *Asia* or *energy*, used 'energy' – in English – as a shorthand for the company's proper name. The different economies of value rendered the word *Asia* incommensurable, and threatened the hegemony and stability of the

ASIA ENERGY

Season's Greetings
and a Happy and Prosperous
New Year 2007

Figure 2. A photocopy of a greeting card circulated by Asia Energy

Journal of the Royal Anthropological Institute (N.S.), 87-107

dominant sign-system. One can call this the semiotic ideology of the protest (Keane 2005). Semiotic ideologies are people's background assumptions about what signs are and how they function in the world. They are concerned not just with signs *per se*, but with the agentive subjects and the acted-upon objects that might be found in the world. Paying heed to the semiotic character of material things also means that the outcome of the signification is not, in principle, settled. This unpredictability creates political possibilities.

The copy of Asia Energy's greeting card with New Year's wishes for 2007 was carefully kept in a senior activist's personal dossier of important paperwork. It came to me about a year later in black-and-white, Xeroxed form. The card features photographs of smiling women and children alongside those of heavy machinery used in mining. It shows power plants presumably fuelled by coal, and snippets of village life complete, as it often is, with a bullock cart on a dirt road. The other side of the card, which is not shown here, has a photograph of a solitary banyan tree – old, majestic, and symbolic of the assumed temporal depth of the relationship between nature and humans and the putative environmental consciousness of the company's policies. Banyan trees in this projected face-to-face world are known for facilitating *communitas* by offering shade, shelter, and sociality. The serialized images on the card formed a predictable tableau of development. Progress is defined by resource use and industrial production. A promise of prosperity, temporalized in the form of anticipation of an auspicious future, tokenizes that which is traditional, and equates what is in the present with an immediate, albeit quaint, past.

In the context of vast illiteracy, the value of the word *Asia* was contingent on the role the card played as an object. Printed on glossy paper, with colour photos and New Year's wishes inscribed in English, the card was a proof of the kinds of secretive exchanges of which the company was increasingly suspected. The greeting card came closest to being an indexical icon (Lee 1997; Silverstein 2003), and pointed to other forms of bribes already mentioned, whose donation and reception were matters of speculation and suspicion. Accepting or declining them – or claims made about both – could align someone with the company or against it. Either could carry significant material consequences for those involved.

Like coal itself, the card's significance relied on the perception of those who saw in the corporation's insignia an index of its misdemeanours, for which it was notoriously unaccountable. They eventually enacted the violence of this 'recognition' on the body of the stranger whose clothing displayed the identical markings. That the trajectory of the card's social life could carry the threat of further violence was reflected in another bureaucratic artefact (cf. Hull 2012). The Phulbari chapter of the National Committee to Protect Oil, Gas, Mineral Resources, Power, and Ports (hereafter the National Committee), a left-leaning citizen's group spearheading the movement, sent a letter to the Emergency government dated 11 January 2007.[7] One of its main demands was the immediate arrest and trial of the *dalal* (agent or traitor) who helped circulate the greeting cards issued by the corporation.

As is obvious, claims to authenticity, political dedication, and expertise were made as much through actions and oratories as by documents. Bruno Latour was not wrong when he termed documents the most despised of all ethnographic subjects (Latour 1988: 54, cited in Riles 2006: 2). For this reason it is perhaps all the more important to pay attention to how much they accomplish in terms of producing cultural knowledge. These graphic artefacts were visual alternatives to coal for those who seemed unable to grasp its value, or make sense of the complex flow chart of mining capital that had at its centre coal rather than other valuables of deeper cultural significance, such as land or gold (Gardner 2012; Hashmi 1992; Sartori 2014).[8] In Phulbari, the uncontrollable movements of these textual 'things' created tensions and rifts between the villagers, the state, the corporation, and the activists. The violence of mining, described in exquisite economy in the writing on the wall, was inseparable from the representational violence where the gaps between intended meanings and their reception had repercussions like the assault on a suspicious stranger.

All that glitters

The trajectory of coal's history as a national resource in Bangladesh reflects and follows a similar fetishization of natural gas. For decades, school children grew up on geography lessons that labelled gas as an abundantly available natural wealth. 'Bangladesh is floating on natural gas' used to be the common refrain. The 1982 World Bank report on the energy sector urged for its effective use and called it a 'crucial element in alleviating the country's current payment problems and enhancing its energy outlook' (cited in Muhammad 2014). At the contemporary level of consumption, the report concluded, the amount of gas was to last for several decades. The general excitement lasted until the 1990s, when it was revealed that no gas would be left by the second decade of the twenty-first century unless new reserves were found. The impending crisis was either delayed or brought closer according to expert reports, supporting Timothy Mitchell's observation that organizing expert knowledge and methods of calculations of natural wealth are deeply politicized processes (Mitchell 2002; see also MacLean 2014). Sheikh Mujibur Rahman, the first President, had allegedly declined selling gas to Gazprom, the Soviet energy company, although the Soviet Union was one of the first countries to recognize Bangladesh. Decades later his daughter, Sheikh Hasina, in her capacity as the head of the polity, made news with her stance on energy. She had famously told US President Bill Clinton on his tour of the country in 2000 that Bangladesh would not consider exporting its most precious natural resource before having met national demands for the next fifty years. Hasina's conversation with Clinton has been a much-publicized nationalist moment that portrayed a brave and farsighted image of the country's leader.

A more careful reading of the energy politics since 1975 (when Rahman was killed in a US-backed coup, following which the military gained political prominence) shows neither foresight nor conviction in the numerous contracts signed between Bangladesh and its foreign partners. Although gas has not been sold to any international party, the reasons are generally ascribed to a lack of consensus within the government on profit sharing or, at the least, electoral considerations. The inability of the Bangladesh government to demand adequate compensation for 'blow-outs' at gas fields managed by Chevron and Niko in the last two decades, no matter which political party has been at the helm, is telling (Gardner 2012). In the absence of a national energy policy, the governmental strategies so far *vis-à-vis* multinational capital have belied the much-hyped Hasina-Clinton exchange.[9] By its decisions to hand over gas blocks to the US energy company Conoco-Phillips for exploration (and possible export) in 2011 and to establish a joint coal-fired power plant with India in Rampal in the southwest, near the world's largest mangrove forest, the Awami League-led government has exposed its apathy towards national and environmental interests around resource management.[10]

Power cuts, as frequent as they were during 2007-8, were thus symptomatic of much that had gone astray in the way of national progress. Despite less than half of the population enjoying access to electricity (Bedi 2015; International Energy Agency 2015), ever-rising consumption created serious lags between demand and supply – a gap of 1,500 megawatts daily, to be precise (Asian Development Bank 2011). The World Bank added to the collective woes by estimating a 10 per cent loss of business sales owing to load shedding (Gunter 2010). Aggravating a much-lamented scarcity, technical, commercial, and political inefficiencies accounted for a waste of nearly 45 per cent of generated power (Gulati & Rao 2007), a trend also noticeable in India and Nepal (Kale 2014; Sovacool, Dhakal, Gippner & Bambawale 2013). In August 2010, a five-hour shut-off of air conditioners and a six-hour close-down of gas stations per day were slapped on the capital city for a month to tackle the peak summer-season crunch (Ethirajan 2010).

It is no surprise, then, that 'failure', 'crisis', 'catastrophe', 'crash landing', and even 'diabolic' were some of the terms used in public discussions of the energy sector. While the suspense in Phulbari rose with Asia Energy contemplating a comeback, larger anxieties about the crisis manifested, one might say, in their most visceral form. In an article titled 'Bangladesh running on empty', its author starts with statistical data to substantiate his prophecy of an imminent disaster. But his narrative quickly switches to an affective register when describing what he calls the 'real' of a dark reality:

> What made the stats 'real' for me was the experience one day last week when we had 10 power cuts in one day in the AT Capital office. Our IPS [Instant Power Supply] backups failed, and I saw 25 young, energetic IBA/BUET [business/engineering] students sitting around doing nothing and feeling dejected and deflated as their computer screens were dead, their backup laptops were also out of power, as were their optimism and enthusiasm (Islam 2008).

The computer screens, ominously dark and dead in the absence of backup power, were signs of a similarly murky national future where young minds from top business and engineering schools sat idly in the absence of work. A grand waste of another kind of energy, no doubt. Human capital has long been a resource that Bangladesh has regularly exported to the developed nations. It is also indispensable to the risk calculus of future labour. Vincanne Adams et al. write:

> Human capacities – to learn, to be healthy, to work, to reproduce – are configured as forms of capital open to speculation not only for individuals and their families, but also for states and transnational investment. Different people from different economic and hence investment milieux are calibrated as having differential rates of return, offering new ways of mapping human possibility (2009: 259).

Half of Bangladesh's national GDP at the time was made up of the service sector, without any meaningful contribution from gas and coal (Gardner 2012). And yet resource-crunches of different kinds, including human capital, got anchored in a presumed crisis of natural resource. 'Bangladesh suffering from the worst ever energy crisis cannot have the luxury of leaving its coal resources buried and opt for LNG [underground coal gasification] and coal import … It is a crime to think of it even', an expert vented in print (Saleque 2011).

To name a material object a resource is to presuppose a historical relationship between humans as exploiters of resources and nature; it is to imagine a fundamental separation between humans, nature, and the resources that emerge in an encounter between the two (Boyer 2011; Ferry & Limbert 2008; Harvey 2001; Latour 1993;

Richardson & Weszkalnys 2014). Scarcity as an idea is built into the act of naming something a resource. If resources are contextually defined, as anthropologists have repeatedly shown (Apter 2005; Strauss, Rupp & Love 2013), then it is more useful to pay attention to 'resource imagination', the social, political, and affective process of ascribing value to objects (Ferry & Limbert 2008). Tanya Richardson and Gisa Weszkalnys introduce the term 'resource environments' to remind us that resource-making is also a material process. Instead of looking at resources as substances with essences that exist naturally, the term guides our attention 'to the complex arrangements of physical stuff, extractive infrastructures, calculative devices, discourses of the market and development, the nation and the corporation, everyday practices ... that allow those substances to exist as resources' (Richardson & Weszkalnys 2014: 7)

Consider Ranjan's words, for example. Ranjan (pseudonym), an elderly farmer, was one of the many locals injured on 26 August 2006. On that day, a siege was organized by the residents of Phulbari and the neighbouring area with the support of the National Committee to demand Asia Energy's departure. The processions from different parts of the potential mining zone started coming in from the early morning, and the approximate number, a few thousand in total (some would say many more; see Luthfa 2011), exceeded the expectations of those in charge. The atmosphere was both festive and anxious; some members of the indigenous communities joined with bows and arrows, cultural groups rendered protest songs, and troops of police and paramilitary guards took position around the township's main thoroughfare and the building rented by Asia Energy. Around 4 p.m., a substantial crowd gathered on the highway close to a bridge a mile or so from the company's office on the other side of the river. The leaders were wrapping up their speeches, delivered from the top of a truck with loudspeakers in hand. The police and the paramilitary border guards, formerly known as the Bangladesh Rifles (BDR), had already barricaded the bridge to stop people from crossing the river to get to the company's office. Just as the leaders were moving away from the barricade behind them, the police began firing rubber bullets and tear-gas shells into the crowd, heightening the tension in the streets. The crowds retaliated by hurling stones while scurrying away from the charging troops. In the ensuing chaos, when some people were already half-way over the bridge and others were crossing the river by wading through the water, the BDR started shooting – first at the sky and then at the scattering crowds.

Ranjan was the first person to brave a bullet. He shared the memory of this attack on his person with considerable excitement. 'I was on top of the bridge ... When I was there, they directly shot at me. I fell on the bridge from the direct shot. I was still alert. Right then I saw blood dripping through my hands. My intestines had come out', he remembered. Though keen on talking about his near-death experience, Ranjan did not seem as thrilled to discuss coal as I was. And he was hardly the only one. Rarely had coal figured in my conversations until I brought it up. As if to please me while presenting himself as an ideologically driven participant in the protests, Ranjan started his answer off on a standard note. Pointing out the unequal royalty arrangements in the contract between Asia Energy and Bangladesh (6 per cent for the host country) and the high-level governmental corruption that would make such an arrangement even imaginable, he stressed the urgency in nationalizing natural resources. Ranjan's response did not stay long within the parameters of a well-rehearsed story, though. He was suspicious of the focus on coal and refused to accept Phulbari's fate as a straightforward victim of an unequal exchange:

Journal of the Royal Anthropological Institute (N.S.), 87-107
© Royal Anthropological Institute 2016

> We cannot be sure that there is nothing more of value (*mulyoban*) here other than coal. We have been seeing very high-quality coal. It shines. When they did the boring, the tests, we went and stood there at the front to watch. They didn't let us in, of course. But you can see it before you as they bring it above ground through the pipe. Then you could see. Now, is it easy to say what else is in it? How could we see? Do you think they'd let us see? They blocked off the area with red ribbons. Their guards did. So that nobody could go in.

Despite the security measures, the value of coal was visible from afar, from where Ranjan stood with the crowd. Phulbari's coal has been known to be of excellent quality by the company as well as various experts and politicians who have been demanding its immediate extraction. The coal here is mostly 'high volatile bituminous' (Asia Energy Corporation 2006), which is of a better quality ('sweet, low ash') compared to metallurgical and thermal coal (Saleque 2011). Publicizing this fact has had political significance. It is widely understood that the coal that Bangladesh regularly imports from its powerful neighbour is of a lesser type (Hussain 2013). This could be – and in fact has been – used as a strong argument for exploiting Phulbari's mines to save foreign currency and curb foreign dependence. The sensitive topic of export, around which discourses of nationalism, energy sovereignty, and anti-imperialism have coalesced, has also often attended to the coal's quality.

For Ranjan, the coal that he saw was of good quality as well, but he appraised it by its visual rather than material value. He described coal not in its absence but in its aesthetic abundance. It was good and glittery. The expression he used to qualify the mineral, *chokchok kora*, to shine, was the same one would use for gold, precious gems, and other objects with surfaces that reflect light.[11] Value creation, Ranjan reminds us, is a complex symbolic process. The experience of qualities is a fact of sociocultural life rather than purported properties of things in the world (Chumley & Harkness 2013; see also Munn 1992). In discussing the perceptions and sensations 'in' and 'of' the mind, or 'qualia', in short, Lily Chumley and Nicholas Harkness (2013) explain that a semiotically informed understanding of qualia as social, rather than purely subjective, reveals how value is constructed through qualitative experience. An ethnography of signs helps situate people's qualitative experiences and their valuations in culture.

There is a slippage here, too, in Ranjan's use of the verb *dekha* – 'to see'. When he said, 'We have been seeing very high-quality coal', he meant seeing in a literal sense. This was the shiny coal that was there for all to see, particularly those who stood on the other side of the guarded area. They saw it when it was brought above ground through the pipes. The second use of *dekha* is of a different, figurative nature. 'Now, is it easy to say what else is in it? Do you think they'd let us see?' – he asked, suggesting other motives behind the extraction of coal, possibly other forms of asset that were more valuable (*mulyoban*) than coal. Whatever other minerals that were also being extracted must have been of a much higher value and therefore kept behind the façade of digging up coal. People like Ranjan could not possibly see this process, as it was not even meant to be seen.

A group of women in a nearby village similarly expressed impatience with my soliciting a discussion about either coal or the mines. 'Doesn't all of Bangladesh have gold mines under it?', the older woman from the group asked dismissively in response to my query. 'What is so precious about coal?' was what she meant by that question, which I would have considered rhetorical had I not had enough experience with others who also approached resources not in terms of scarcity or lack. This is different from what Katy Gardner (2012) has heard around the gas fields in Bangladesh's northeast.

'Under our fields is gold', a local woman says to the anthropologist. Gardner gives this as an example of the knowledge that the residents of this area possessed of the value of the wealth in their proximity. 'Savvy entrepreneurs that the locals are', she writes, 'everyone is aware of the economic potential of the gas that the foreigners discovered' (Gardner 2012: 3). My argument here is that while people in Phulbari were made aware of the heightened valuation of the resource underneath, they were not convinced why coal (as opposed to other resources) was valued as such. They were more curious as to what all this official interest *actually* signified.

Popular thinking about coal, even when inflected by anti-mining rhetoric, revealed an understanding of resource that was different, if not antithetical, to a conception of it as scarce that is common between the resistance movement, the company, and its ally, the Bangladeshi state. Scarcity here was countered by both qualitative and quantitative excess. This was further pointed out to me when others talked about digging out coal from their portion of agricultural land if the country needed it, 'just like we plough our land for crops'. Making coal visible, or domesticating it in this way, has been one strategy to counter the signs of a dim future made even murkier by the threat of violence and displacement.

Dark futures (or just 'eyewash?')

If Phulbari's uprising interrupted the official dreams of an electrified future, the state of emergency of 2007-8 inaugurated a different kind of anticipation. On 11 January 2007, the President dissolved the parliament, initiating an official state of emergency that lasted until December 2008. Its current invocation was novel even in a country where multiple coups and martial laws have taken place regularly since independence (Muhammad 2008; Umar 1980). By late 2006, the topic of criminalization of politics had become more than a national pastime. This was particularly so as a failed dialogue between the two largest political parties regarding the terms of the national elections ended in fatalities. The Bangladesh Awami League (BAL) and the Bangladesh Nationalist Party (BNP) have dominated national politics since the latter's founding in 1978. The League has a history longer than its arch-rival, having been born of the All-Pakistan Awami Muslim League, formed in 1949 in Dhaka, the East Pakistan capital. Both have alternatively formed the ruling majority since 1990, headed by the daughter and the widow of former presidents, Sheikh Mujibur Rahman and Ziaur Rahman, respectively. In 2006, the parties were once more at loggerheads on the nitty-gritty of the national elections and the design of the caretaker government. Since the fall of dictatorship in 1990, the interim government, generally headed by a non-partisan member of the civil society, has been a constitutionally sanctioned solution to the problem of governmental complicity in electoral fraud.

For the most part, the official view of the Emergency as a necessary 'disturbance' to the democratic process found popular, albeit short-lived, support (Chowdhury 2014). The efforts to cleanse the culture of politics were left to the discretion of foreign diplomats, local technocrats, and the army (bdnews24.com 2011), an alliance since described as Bangladesh's 'civic-military-corporate' democracy (Wasif 2009). The anti-corruption programme of the interim government focused its energy primarily on accounting for untaxed income, or 'black money'. It also attempted serious aesthetic airbrushing of the body politic. Slums, the archetypal eyesore on South Asia's cityscapes, were forcefully evicted within hours of its declaration as a first step towards beautifying the major cities. Transparency, as seen in the Indian Emergency in the mid-1970s, was the operative ideology of this regime (Tarlo 2003). The term privileges the visual register. And yet local visions of the contemporary moment challenge or are deeply suspicious of the touted 'openness' and 'accountability' of the transparency discourse (West & Sanders 2003). International financial and development organizations demand transparency from poorer nations in exchange for funds and political support, while ordinary people observe modernity's consequences through a critical lens that is often opaque (Hetherington 2012; West & Sanders 2003). In Bangladesh, a democratic future where violence would be a thing of the past – promised, ironically, by a military-backed regime – looked different from the vantage of Phulbari's everyday life.

The conditions of my fieldwork were affected by the exaggerated nature of state surveillance. As a pre-emptive measure to avoid harassment or even potential expulsion from Phulbari in the name of 'personal security', I thought of introducing myself to the local authorities. The decision met with apprehension from friends and allies well acquainted with the local scene. The potential futility of my effort at establishing documentary legitimacy was caricatured in the story I was told as a reaction to my suggestion. The activist with a top-level position in anti-mining leadership asked me if I was familiar with Krishan Chander's writing. Chander was a well-known progressive Hindi-Urdu fiction writer of the mid-nineteenth century whose work has been translated into many South Asian languages, including Bengali. The story I was told was entitled 'Chapapora manush' ('The man under the tree'[12]). My friend's version of Chander's short story went something like the following:

> A man was stuck under a felled tree during a storm. When others came to help him out from under the tree, they noticed that the tree that had trampled the victim was no ordinary tree; it belonged to the Forest Department. It was this government office that had to be notified of the accident before any measure could be taken to save the man. Upon being notified of the event, and following the established chain of command, the Forest Department informed the police, who felt obliged to report to the Home Ministry. Ultimately the Prime Minister had to be made cognizant of the accident. After all, it involved a human being. At the end, the person who cut the branches to help relieve the man of the weight of the tree was greeted simply with a skeleton.

'So, this is how the Bangladeshi administration works', my interlocutor said following the story's morbid punchline. 'By the time your information gets to the [office of the] District Commissioner, they might seriously wonder if you are a representative of Tata [the Indian business giant] or if you are a spy', he explained, smiling. At this time Tata was showing a keen interest in exploiting the gas reserves in the northeast and in founding a steel plant in Bangladesh. A potential customer of Phulbari's coal, it was suspected of powerful political machinations within the government the same way Asia Energy was. The satirical impulse of my friend's story may not be so alien to those with experiences with the bureaucratic machinery in South Asia (Gupta 2012; Hull 2008; Visvanathan & Sethi 1998). A veritable labyrinth, bureaucracy needs to be navigated strategically to get anything done, even when what may be in order is an urgent response to save a human life. Our narrator's smile was a knowing one, generated as much from contentment in possessing knowledge of how the state *really* worked as from satisfaction in being able to show the naïve outsider the ropes, so to speak.

The pleasure of the storyteller was partly derived from his narrative's putative shock value. After all, death and other forms of state-sponsored violence were dangerously close to Phulbari's reality. People got randomly killed when protesting against the corporation; a high-ranking local activist was publicly humiliated, beaten, and held up by the military on account of fabricated charges of theft; a young Internet activist and the founder-moderator of a web forum on Phulbari was interrogated by members of Rapid Action Battalion, the elite police force. They 'suggested', after a relatively soft display of might, that young people like him should go abroad for Ph.D.s and not get their hands muddied in such anti-nationalistic initiatives as stopping coal mines, which could discourage foreign direct investment.[13] Power worked not in unpredictable and capricious ways but in a precise and exacting fashion, its banality and absurdity notwithstanding.

The violence inherent in bureaucratic routinization, of which the state is an exemplar, was made visible in the story about the death. The corpse as the final evidence was a sign of the intricacies of power, whose physical locations are never easily traceable. 'I die, therefore I am', the subject of stately being, crushed under the government-owned tree, may very well be saying (Taussig 1997). The material remains of the man was a visual shorthand for the circuits of power that worked through a strategy that many people described to me as 'eyewash'. Eyewash, used in English when speaking or writing in Bengali, is best understood as the conceptual other of transparency. It challenged some of the official signs of democracy's future in Bangladesh.

In 2006, the government of Bangladesh declined a much talked-about Policy Support Instruments (PSI) treaty initiated by the International Monetary Fund (IMF). This rare resistance to a powerful donor organization surprised many observers, including some of my friends in Phulbari. 'Eyewash' was on offer as a ready explanation.[14] My activist friend had another story to share in relation to the news that made the headlines that day. This time it was an anecdote from his life. Well into his forties at the time of our conversation, the leader was remembering an event from childhood. When he was a young boy, a few older boys from his village had a picnic. The food was taking a long time to prepare. To make matters worse, a couple of the older boys got into a serious brawl, which scared off a few guests, who left without eating. Later in the evening, when our storyteller went back with a friend to see what was happening, not happy to have left the picnic spot hungry, they found the older boys enjoying the meal together. The fight was only a trick – 'eyewash', he said in English – to get rid of the unwanted guests. The government of Bangladesh was doing exactly the same, he said, by way of concluding the story.

There could hardly be a more apt figuration of the national dramaturgy than in this story's main plot: that of two older boys staging a fight as a ploy to fool hordes of young children. Was not an army-backed caretaker government offered as the single viable alternative when the senior representatives of the Bangladesh Awami League and the Bangladesh Nationalist Party failed to reach a consensus? All hell broke loose when party members started fighting in the streets, using the breakdown of the dialogue as an excuse for violence. In retrospect, the mass-mediated performances of deliberation between the two general secretaries were widely read as a sham effort at consensus-building. An influential political leader in Phulbari spelled it out for me. He was one of the rare few who supported and contributed to the protest movement and still maintained strong allegiances to a mainstream political camp.

> The people of our country need to be alert. What is going on right now is *eyewash* . . . This *village politics* have started since Mannan Bhuiyan and [Abdul] Jalil [the general secretaries of the two parties] have started sitting down together. The foreigners have *incited* both the parties. The people of the country then became divided. They lost faith in the two leaders. [Italicized words are in English in the original.]

Journal of the Royal Anthropological Institute (N.S.), 87-107
© Royal Anthropological Institute 2016

To describe eyewash as a strategy and outcome of 'village politics' is to parochialize the careful deceptions that were popularly read into the woeful charade of a national political dialogue, or the government's decision not to pay heed to the IMF. It is curious, though, that village politics, with its attendant connotations of petty, small-minded and provincial, was used to denigrate the incitement of foreigners, in this case the powerful diplomatic envoys from the United States, the United Kingdom, and the European Union. Their decisive role in changing the course of national politics had not gone completely unnoticed even in this peripheral township. Eyewash did not merely signal that power was opaque; it also asserted a certainty about the knowledge of how it functioned that countered the anticipatory regime of the state. Eyewash was a popular tactic to look past the signs of development and democracy made abundantly available by the Bangladeshi state and its foreign donors and allies, and a nationally and internationally sustained ethic of probability (Appadurai 2013).

A few years later when the WikiLeaks reports hit the news, Global Coal Management Resource's investors' forum was abuzz with rumours, conjectures, and anticipation. On-line the investors deliberated on the possible impact of these revelations of powerful foreign connections on Phulbari and on Bangladesh, and on their own privatized financial futures. Laced with ethnocentrism, the comments sent from around the globe re-signified the notion of parochial politics, or village politics, as the leader from Phulbari described it. 'I think we will be OK as most of the people [in Bangladesh] can't read English . . . Great to read there is pressure at that level', an optimistic investor reacted to the documentary proof of top-level connections exposed in the cable leak.[15] The particular investor as a part of a collective 'we' voiced a hope that rested on a possible misrecognition. A potential inability to comprehend signs – *most of the people can't read English* – was conducive to a lucrative future. Yet the comment also spoke of an anxiety about what might happen if Bangladeshis actually understood these correspondences in English and thwarted the efforts to sustain the structural violence that many of them experienced on a daily basis. Phulbari's ghost may very well be haunting the desires virtually expressed on this resource-orientated investment site.

It is clear that anticipatory regimes pervade the ways we think about, feel, and address our contemporary problems (Appadurai 2013; Koselleck 2004; Massumi 2007). As Vincanne Adams *et al.* further note, 'Promissory capital speculation and development logics render some places as backward in time, needing anticipatory investment, while other places are deemed already at the cusp of the "new" future, marked by the virtue of rapid change' (2009: 251). While I agree that anticipation is a critical element in contemporary global politics, in light of the observations presented here, I follow Arjun Appadurai in thinking of the future as a cultural fact; it is not a blank space for the inscription of technocratic enlightenment, but a space for democratic design (Appadurai 2013: 299). The everyday negotiations between the ethics of possibility and the ethics of probability involve multiple readings of signs. The miscomprehensions that arise from them are always political. The unintended slippages in meanings, the inadvertent confusion of words, and the consequent hazards of representation (Keane 1997) can be, and often are, significant sources – or resources – of political action. In the explosive political space of Phulbari, the textual and the mineral were both invaluable resources, though in disparate forms and for different people. Popular efforts to make certain futures visible surfaced in and through these slippages, whether they were marked on the body of a suspicious outsider or were etched on the wall of a mud house.

Journal of the Royal Anthropological Institute (N.S.), 87-107
© Royal Anthropological Institute 2016

NOTES

An Overseas Dissertation Research Grant (Dean's Office) and the Committee on Southern Asian Studies Dissertation Grant at the University of Chicago, together with a Charlotte W. Newcombe Doctoral Dissertation Fellowship from the Woodrow Wilson National Fellowship Foundation, generously funded fieldwork in Bangladesh and the subsequent write-up. I am grateful to William Mazzarella and Joseph Masco for their comments on earlier versions of this essay. It has also benefitted from the discussions at the Global Environments Workshop at the University of Chicago, 2011; the 'Contesting Pasts, Competing Futures' workshop on Bangladesh at the University of Texas at Austin, 2013; and the American Ethnological Society annual meeting panel, 'Visibilization and Concealment: Social Critique and Anthropologies of Value', in Boston, 2014. I want to extend my warmest thanks to Jessica Barnes for inviting me to contribute to the special issue. I thank the anonymous reviewers for *JRAI* for the final version you read here.

[1] The gap between everyday speech and the standard used in newspaper and other mass media was overcome in this spelling mistake, which, in the process, generated a clever and powerful pun. Despite the numerous texts produced for domestic and international consumption, this unintentionally humorous substitution was not adopted as a potent tagline of the movement. This was because it was not written in the elevated register of standard Bengali, the preferred medium for discourses of expert knowledge and formalized activism.

[2] Asia Energy Corporation (Pty) Ltd. Press release, Dhaka, 21 June 2006.

[3] I hesitate to give an exact date because of the confusion around the issue of acquiring licensing for mining. Starting with BHP, an Australian company that transferred its licence to Asia Energy, which soon became a subsidiary of Global Coal Management and, even later, GCM, Plc, the story of signing the contract between the government of Bangladesh and the corporation is highly contested. Over the years, both (in their various incarnations) have retracted or revised earlier versions of how and when the coal project was given an official nod, if at all, and under what conditions.

[4] Despite this statement, as Christine Haigh of Global Justice Now notes, NCP's report did not take into account the possible human rights violations should the mining actually take place. Also, the note in the report that GCM has failed to 'foster confidence and trust in the society in which it operates' is a gross understatement given the number of deaths, injuries, and other forms of violence that had already occurred in Phulbari (Haigh 2015).

[5] Although the first of its kind in Bangladesh, the form and content of the Phubari movement shares many similarities with the events in Nandigram in neighbouring West Bengal (India). A few months after the violence in Phulbari, in March 2007, farmers in Nandigram challenged the state acquisition of land for a chemical hub. The resulting violence led to at least fourteen deaths. (For more details on the political and social context of Nandigram, see Cross 2014; Dhara 2015.)

[6] The gifts distributed among the villagers over time became suspect because of the notorious corruption of Asia Energy. Despite some early meetings in 2005 when the company tried to create awareness of the project, their activities were nothing compared to the elaborate corporate social responsibility activities noted, for example, by Marina Welker (2009) in Indonesia. The lack of information available in Bengali and the misrepresentation of the villagers' presence at these meetings as their 'approval' for the project were deeply resented by the residents.

[7] A non-governmental group, the National Committee was founded in 1998 to address and resist national resource extraction by multinational capital. Its founding members were largely ideologically left, and included the participants of Baam Ganotantrik Front (Left Democratic Front) and a consortium of eleven smaller and more radical left parties (Shahidullah n.d.). It has been a relatively diverse organization with the active involvement of intellectuals, feminist leaders, and student activists. Its Phulbari chapter was formed in 2005 and has since aided in organizing and publicizing events around the coal project and sustaining the protest culture.

[8] The value generally attached to land in peasant societies across time and space also gets projected onto documents, such as titles, deeds, and signboards. A common index of proprietorship across South Asia are the signboards that are physically dug into the land to ward off possible claimants of ownership by signifying legal sanction.

[9] 'US, Bangladesh share commitment to peace, democracy, free markets' is the subtitle of the official transcript of the Clinton-Hasina joint press statement published on the US embassy website.

[10] In September 2015, Sheikh Hasina won the United Nations Champions of the Earth award for Bangladesh's investment in fighting climate change. Though welcomed by many groups fighting for the environment within Bangladesh, this development also prompted fresh demands for the government to

scrap the Rampal Power Plant that directly threatens the ecology of the Sundarban mangrove forest (United Nations News Service 2015).

[11] Indeed, the common Bengali translation of the English proverb 'All that glitters is not gold' is '*chokchok korilei sona hoyna*', which has the same verb that was used by Ranjan.

[12] My translation.

[13] E-mail sent to Phulbari_Resistance listserve in April 2008.

[14] For instance, the Phulbari Treaty signed between the people of Phulbari and the government in 2006 was described in some media as an 'eyewash'. Among other things, the treaty demanded the expulsion of Asia Energy, a ban on the export of coal and the involvement of foreign companies, and compensation for the dead and wounded. The Emergency government questioned the authenticity of the document. 'The promise to void the contract signed with Asia Energy made by the BNP [Bangladesh Nationalist Party] and Jamaat [e Islami] alliance government was an eyewash; the promise was made to divert the attention of the protesting people', claimed an article published in August 2007, precisely a year after the signing of the treaty (*Financial Express* 2007).

[15] Quoted from GCM's on-line discussion board, which was cited in an e-mail sent to Phulbari_Action listserve on 22 December 2010.

REFERENCES

ADAMS, V., M. MURPHY & A.E. CLARKE 2009. Anticipation: technoscience, life, affect, temporality. *Subjectivity* **28**, 246-65.

APPADURAI, A. 2013. *The future as cultural fact: essays on the global condition.* London: Verso.

APTER, A.H. 2005. *The Pan-African nation: oil and the spectacle of culture in Nigeria.* Chicago: University Press.

ASIA ENERGY CORPORATION 2006. Bangladesh: Phulbari Coal Project. Summary environmental impact assessment. Prepared by Asia Energy Corporation (Bangladesh) Pty, Ltd for the Asian Development Bank (ADB).

ASIAN DEVELOPMENT BANK 2011. Speech by Paul Heytens, Country Director, Asian Development Bank in Dhaka at the ICC Conference on Energy and Growth. Asian Development Bank.

BDNEWS24.COM 2011. WikiLeaks exposé: Bangladesh. Int'l help for military solution was sought. bdnews24.com, 6 September (available on-line: *http://bdnews24.com/bangladesh/2011/09/04/wikileaks-exposc-bangladeshint-l-help-for-military-solution-was-sought*, accessed 8 January 2016).

BEDI, H.P. 2015. Right to food, right to mine? Competing human rights claims in Bangladesh. *Geoforum* **59**, 248-57.

BOYER, D. 2011. Energopolitics and the anthropology of energy. *Anthropology News* **52**: 5, 5-7.

CHOWDHURY, N.S. 2014. 'Picture-thinking': sovereignty and citizenship in Bangladesh. *Anthropological Quarterly* **4**, 1257-78.

CHUMLEY, L.H. & N. HARKNESS 2013. Introduction: qualia. *Anthropological Theory* **13**, 3-11.

CROSS, J. 2014. *Dream zones: anticipating capitalism and development in India.* London: Pluto.

DHARA, T. 2015. Nandigram revisited: the scars of battle. Infochange (available on-line: *http://infochangeindia.org/agenda/battles-over-land/nandigram-revisited-the-scars-of-battle.html*, accessed 8 January 2016).

ETHIRAJAN, A. 2010. Bangladesh to shut gas stations amid power crisis. BBC, 12 August, sec. South Asia (available on-line: *http://www.bbc.co.uk/news/world-south-asia-10960337*, accessed 8 January 2016).

FE INVESTEGATE 2007. Asia Energy PLC announcements. Asia Energy PLC: change of name. Investegate.co.uk, 17 January (available on-line: *http://www.investegate.co.uk/article.aspx?id=200701111034383583P*, accessed 8 January 2016).

FERRY, E.E. & M.E. LIMBERT (eds) 2008. *Timely assets: the politics of resources and their temporalities* (School for Advanced Research Advanced Seminar Series). Santa Fe, N.M.: School for Advanced Research Press.

FINANCIAL EXPRESS 2007. Minu's deal on Phulbari Coal Mine illegal: Law Ministry. 30 August.

GARDNER, K. 2012. *Discordant development: global capitalism and the struggle for connection in Bangladesh.* London: Pluto.

GIDWANI, V. 2008. *Capital, interrupted: agrarian development and the politics of work in India.* Minneapolis: University of Minnesota Press.

GULATI, M. & M.Y. RAO 2007. Corruption in the electricity sector: a pervasive scourge. In *The many faces of corruption: tracking vulnerabilities at the sector level* (eds) J.E. Campos & S. Pradhan, 115-58. Washington, D.C.: World Bank.

Journal of the Royal Anthropological Institute (N.S.), 87-107
© Royal Anthropological Institute 2016

GUNTER, B.G. 2010. The impact of development on CO_2 emissions: a case study of Bangladesh until 2050. Falls Church, Va: Bangladesh Development Research Center (BDRC) (available on-line: *http://www.bangladeshstudies.org/files/WPS_no10.pdf*, accessed 8 January 2016).

GUPTA, A. 2012. *Red tape: bureaucracy, structural violence, and poverty in India.* Durham, N.C.: Duke University Press.

HAIGH, C. 2015. The global system for holding corporations to account is in need of serious reform. *Guardian*, 10 February (available on-line: *http://www.theguardian.com/global-development-professionals-network/2015/feb/10/the-global-system-for-holding-corporations-to-account-is-in-need-of-serious-reform*, accessed 8 January 2016).

HARVEY, D. 2001. *Spaces of capital: towards a critical geography.* New York: Routledge.

HASHMI, T. U.-I. 1992. *Pakistan as a peasant utopia: the communalization of class politics in East Bengal, 1920-1947.* Boulder, Colo.: Westview.

HETHERINGTON, K. 2012. Agency, scale, and the ethnography of transparency. *PoLAR: Political and Legal Anthropology Review* **35**, 242-7.

HULL, M.S. 2008. Ruled by records: the expropriation of land and the misappropriation of lists in Islamabad. *American Ethnologist* **35**, 501-18.

——— 2012. Documents and bureaucracy. *Annual Review of Anthropology* **41**, 251-67.

HUSSAIN, D. 2013. *Boundaries undermined: the ruins of progress on the Bangladesh/India border.* London: C. Hurst & Co.

INTERNATIONAL ENERGY AGENCY 2015. International Energy Agency: Bangladesh – indicators for 2012.

ISLAM, I. 2008. Bangladesh running on empty? *Daily Star*, 10 April (available on-line: *http://www.thedailystar.net/story.php?nid=31485*, accessed 8 Janury 2016).

KALAFUT, J. & R. MOODY 2008. *Phulbari Coal Project: studies on displacement, resettlement, environmental and social impact.* Dhaka: Samhati Publications.

KALE, S. 2014. *Electrifying India: regional political economies of development.* Stanford: University Press.

KARIM, F. 2010. WikiLeaks cables: US pushed for reopening of Bangladesh coal mine. *Guardian*, 21 December (available on-line: *http://www.guardian.co.uk/world/2010/dec/21/wikileaks-cables-us-bangladesh-coal-mine?INTCMP=SRCH*, accessed 8 January 2016).

KEANE, W. 1997. *Signs of recognition: powers and hazards of representation in an Indonesian society.* Berkeley: University of California Press.

——— 2005. Signs are not the garb of meaning: on the social analysis of material things. In *Materiality* (ed.) D. Miller, 182-205. Durham, N.C.: Duke University Press.

KHAN, T.M. 2006. Mineral resources, Phulbari movement and lessons from Nigeria. *Daily Star*, 17 September (available on-line: *http://archive.thedailystar.net/2006/09/12/d609121501122.htm*, accessed 8 Janury 2016)

KIRSCH, S. 2014a. *Mining capitalism: the relationship between corporations and their critics.* Berkeley: University of California Press.

——— 2014b. Imagining corporate personhood. *PoLAR: Political and Legal Anthropology Review* **37**, 207-17.

KOSELLECK, R. 2004. *Futures past: on the semantics of historical time* (trans. K. Tribe). New York: Columbia University Press.

LATOUR, B. 1988. Drawing things together. In *Representation in scientific practice* (eds) M. Lynch & S. Woolgar, 19-68. Cambridge, Mass.: MIT Press.

——— 1993. *We have never been modern* (trans. C. Porter). Cambridge, Mass.: Harvard University Press.

LEE, B. 1997. *Talking heads: language, metalanguage, and the semiotics of subjectivity.* Durham, N.C.: Duke University Press.

LUTHFA, S. 2011. Everything changed after the 26th: repression and resilience against the Phulbari Coal Mine, Bangladesh. Working Paper 193. Queen Elizabeth House Working Paper Series. University of Oxford (available on-line: *https://www.academia.edu/1013131/Everything_Changed_after_the_26th_Repression_and_Resilience_against_the_Phulbari_Coal_Mine_Bangladesh*, accessed 8 Janury 2016).

MACLEAN, K. 2014. Counter-accounting with invisible data: the struggle for transparency in Myanmar's energy sector. *PoLAR: Political and Legal Anthropology Review* **37**, 10-28.

MASSUMI, B. 2007. Potential politics and the primacy of preemption. *Theory and Event* **10**: 2.

MITCHELL, T. 2002. *Rule of experts: Egypt, techno-politics, modernity.* Berkeley: University of California Press.

MORRIS, R.C. 2008. Rush/panic/rush: speculations on the value of life and death in South Africa's age of AIDS. *Public Culture* **20**, 199-231.

MUHAMMAD, A. 2007. *Phulbari, Kansat, Garments 2006.* Dhaka: Srabon Prokashoni.

——— 2008. Dui dharar rajnitee, totwabodhayok sarkar o nirbachan [Bi-partisan politics, caretaker government, and elections]. In *Kothay jachhe Bangladesh?* [Where is Bangladesh heading?], 167-83. Dhaka: Samhati Prokashon.

Journal of the Royal Anthropological Institute (N.S.), 87-107
© Royal Anthropological Institute 2016

————— 2014. Natural resources and energy security: challenging the 'resource-curse' model in Bangladesh. *Economic and Political Weekly*, 25 January (available on-line: *https://www.academia.edu/5782747/Natural_Resources_and_Energy_Security_Challenging_the_Resource-Curse_Model_in_Bangladesh*, accessed 8 January 2016).

————— 2015. Bangladesh – a model of neoliberalism: the case of microfinance and NGOs. *Monthly Review*, March (available on-line: *http://monthlyreview.org/2015/03/01/bangladesh-a-model-of-neoliberalism/*, accessed 8 January 2016).

Muller, M. & R. Moody 2009. Bangladesh's untapped coal potential. *Daily Star*, 2 June (available on-line: *http://archive.thedailystar.net/newDesign/news-details.php?nid=90796*, accessed 8 January 2016).

Munn, N.D. 1992. *The fame of Gawa: a symbolic study of value transformation in a Massim society* (New edition). Durham, N.C.: Duke University Press.

Papadopoulos, D., N. Stephenson & V. Tsianos 2008. *Escape routes: control and subversion in the twenty-first century*. London: Pluto.

Richardson, T. & G. Weszkalnys 2014. Introduction: resource materialities. *Anthropological Quarterly* **87**, 5-30.

Riles, A. 2006. Introduction: in response. In *Documents: artifacts of modern knowledge* (ed.) A. Riles, 1-38. Ann Arbor: University of Michigan Press.

Saleque, E.K.A. 2011. Bangladesh's energy crisis: anatomy of failure. *Energy Bangla*, 9 May.

Sartori, A.S. 2014. *Liberalism in empire: an alternative history*. Oakland: University of California Press.

Shah, A. 2010. *In the shadows of the state: indigenous politics, environmentalism, and insurgency in Jharkhand, India*. Durham, N.C.: Duke University Press.

Shahidullah, P.S.M. n.d. Bangladesher tel-gyas-khonij sampad o jatiyo swartho rokkhar andoloner sankhipto itibritto [A short history of the movement to protect oil, gas, mineral resources, and national interests of Bangladesh].

Shever, E. 2012. *Resources for reform: oil and neoliberalism in Argentina*. Stanford: University Press.

Silverstein, M. 2003. Indexical order and the dialectics of sociolinguistic life. *Language & Communication, Words and Beyond: Linguistic and Semiotic Studies of Sociocultural Order* **23**, 193-229.

Sobhan, R. (ed.) 2005. *Privatization in Bangladesh: an agenda in search of a policy*. Dhaka: Center for Policy Dialogue/The University Press Ltd.

Sovacool, B.K., S. Dhakal, O. Gippner & M.J. Bambawale 2013. Peeling the energy pickle: expert perceptions on overcoming Nepal's electricity crisis. *South Asia: Journal of South Asian Studies* **36**, 496-519.

Strauss, S., S. Rupp & T. Love (eds) 2013. *Cultures of energy: power, practices, technologies*. Walnut Creek, Calif.: Left Coast Press.

Tarlo, E. 2003. *Unsettling memories: narratives of the Emergency in Delhi*. Berkeley: University of California Press.

Taussig, M.T. 1997. *The magic of the state*. New York: Routledge.

Tsing, A.L. 2005. *Friction: an ethnography of global connection*. Princeton: University Press.

Umar, B. 1980. *Towards the Emergency*. Dacca: Muktadhara.

United Nations News Service 2015. UN News – Bangladeshi Prime Minister wins UN Environment Prize for leadership on climate change. UN News Service Section, 14 September (available on-line: *http://www.un.org/apps/news/story.asp?NewsID=51865#.VgohprRViko*, accessed 8 January 2016).

Visvanathan, S. & H. Sethi 1998. *Foul play: chronicles of corruption*. New Delhi: Banyan Books.

Wasif, F. 2009. *Joruri obosthar amolnama: Bangladesher civiko-military-corporate ganatantra* [Verdict on a state of emergency: Bangladesh's civic-military-corporate democracy]. Dhaka: Shuddhashar.

Welker, M.A. 2009. 'Corporate security begins in the community': mining, the corporate social responsibility industry, and environmental advocacy in Indonesia. *Cultural Anthropology* **24**, 142-79.

West, H.G. & T. Sanders (eds) 2003. *Transparency and conspiracy: ethnographies of suspicion in the New World Order*. Durham, N.C.: Duke University Press.

Witsoe, J. 2013. *Democracy against development: lower-caste politics and political modernity in postcolonial India*. Chicago: University Press.

Mines et signes : ressources et futurs politiques au Bangladesh

Résumé

En 2007-2008, les spéculations sur une catastrophe énergétique allaient bon train au Bangladesh, parallèlement à des visions utopiques de la démocratie. Deux événements sont venus renforcer à la fois le désespoir et l'espoir d'un futur national collectif : un mouvement d'opposition au projet de mines

Journal of the Royal Anthropological Institute (N.S.), 87-107
© Royal Anthropological Institute 2016

de charbon de Phulbari, dans le nord-ouest, et une urgence politique nationale. Le présent article leur est consacré. En croisant une crise sur les ressources et une crise politique, l'auteure affirme, d'une part, que les ressources et la démocratie sont devenus deux locus exemplaires de réflexion sur l'avenir, et d'autre part que les lacunes dans leurs différents imaginaires sont des espaces potentiels du politique. A partir d'une ethnographie des protestations contre la mine de Phulbari, elle analyse la relation entre signification et anticipation dans la politique bangladaise. Elle identifie ainsi les mouvements politiques émergents qui remettent en cause la politique d'anticipation de l'État et de la société de production d'énergie.

6

Chronicle of a disaster foretold: scientific risk assessment, public participation, and the politics of imperilment in Bristol Bay, Alaska

KAREN HÉBERT *Yale University*

Like many environmental controversies today, the debate over the proposed Pebble Mine in a salmon-producing region of Alaska centres on the development and contestation of scientific projections of risk. This paper traces the participatory public process surrounding a risk assessment of potential mining impacts to examine how forums that join expert and lay knowledge shape scenarios of future imperilment and influence environmental politics in the present. It draws on ethnographic research to analyse how risk assessments demand the delineation of spatial, social, and temporal boundaries that provoke tensions, or 'overflows', which reveal the constraints of existing frameworks. In the Pebble debate, the public process generated overflows that expose conflicting claims to knowledge and authority, reflecting the risk assessment's overarching, if often frustrated, effort to separate scientific and technical truths from political contestations. The paper shows how these overflows spurred generative effects, new visions that remake spatial, social, and temporal relations in the face of imperilment. It argues that despite the limitations of common consultative processes and discourses of risk, the negotiation of multiple forms of knowledge and authority in the public view can nevertheless open new spaces and social formations for the exercise of politics.

'I don't think that the mine will ruin our Bristol Bay life', a young woman named Jasmine Kritz stated for the record at an August 2014 public hearing. 'I think it will completely destroy it' (EPA 2014*d*: 111-12). Before a packed room at the high school gym in Dillingham, Alaska, audience members seated on folding chairs and bleachers got up one by one to offer testimony to visiting representatives of the US Environmental Protection Agency (EPA). The EPA had come to this community of 2,500 people in the southwest Alaskan region of Bristol Bay, a remote area known for its robust wild salmon fisheries, to solicit public comment on its proposal to block the development of a controversial proposed mine, the Pebble Mine. The EPA's possible action followed from a major scientific study it conducted, known as the Watershed Assessment, which evaluated the risks of large-scale mining to Bristol Bay's salmon ecosystems.

Authored by EPA officials and independent consultants, the Watershed Assessment draws heavily on existing scientific research along with extensive public input. It 'uses the

Journal of the Royal Anthropological Institute (N.S.), 108-126
© Royal Anthropological Institute 2016

well-established methodology of an ecological risk assessment', the report states, 'which is a type of scientific investigation that provides technical information and analyses to foster public understanding and inform future decision making' (EPA 2014*a*: 1). The study incorporated vehicles for public engagement at nearly every stage, including multiple rounds of public hearings, comment periods on report drafts, and dozens of meetings by EPA officials with key stakeholders in Bristol Bay communities, Anchorage, Seattle, and Washington, D.C.[1] By the time of the hearing in the Dillingham gym, such consultative exercises were a familiar ritual in Bristol Bay. 'As it stands now', Dillingham resident Robyn Chaney observed in her comments, 'my youngest three children have only known hearings and meetings and more meetings and the threat of Pebble Mine' (EPA 2014*d*: 59).

Structured through a series of probabilistic risk scenarios, the Watershed Assessment's 2014 final report presents a worrying picture of prospective mining impacts. Its images of toxic tailings breaches, acid mine drainage, and contaminated water tables reappeared frequently in public testimony. Many Bristol Bay residents referenced calculations found in the Watershed Assessment to spell disaster for a vibrant watershed should the Pebble Mine go forward. Others probed these details to contest forecasts of destruction. On either side of the debate, those presenting public comments underscored the implications for the region's people, most of whom identify as Alaska Native. 'The short life of a sulfide mine [is] 50-70 years and maybe a little longer, 75 years', Sergie Chukwak testified,

> but once that activity is done the mine is just going to sit there with all of the chemicals and continue to kill off more of the rivers and streams, and then instead of just 24% of the Nushagak and the Kvichak river contaminated it is going to be 100% and that is going to be death of 37 communities in Bristol Bay (Chukwak in EPA 2012*c*: 6).

As such comments suggest, Bristol Bay residents employed concepts of scientific risk to advance multi-layered concerns about the future of the region in light of possible mineral development.

Testimony from the public hearings provides a sense of how the prospective mine takes shape as a risk through modes of participation that are characteristic of environmental governance today. In this paper, I examine the public process surrounding the risk assessment in Bristol Bay to analyse how visions of ecological imperilment come into being, and how they influence environmental politics in the present. What does invoking the future through forecasts of harm mean for the people and places that are positioned as threatened? How do mechanisms of public participation affect the formation of disaster prognoses given the negotiation of different forms of knowledge and authority they entail? And what does this suggest about the potential for participatory approaches to risk assessment to promote more meaningfully democratic environmental decision-making on matters of long-term concern?

As the Pebble debate has intensified over the past decade, it has come to hinge ever more prominently on scientific analyses of risk that are not simply communicated to the public but are actually formed through public engagement. In the recurring 'hearings and meetings and more meetings' that have marked the EPA's work in Bristol Bay, the legitimacy of the process is derived through the involvement of both experts and laypersons. Bristol Bay residents have been called upon to put forth their views and visions, though typically with reference to the parameters that structure scientific risk

Journal of the Royal Anthropological Institute (N.S.), 108-126
© Royal Anthropological Institute 2016

assessment. This persistent feature of the participatory project raises questions about the extent to which efforts to engage laypersons on issues of risk, however well intentioned and successfully executed, offer genuine opportunities for negotiating different modes of knowledge and authority to arrive at outcomes that are not already predetermined by expert designs.

My analysis demonstrates that while the public process initiated by the EPA succeeds in opening the production of scientific knowledge to a wider set of actors than typically play a role in its making, it also reinscribes the validity of some forms of authority over others and limits the political possibilities of certain spokespersons. To account for these findings, I take up Brian Wynne's (2002; 2005) suggestion that the idiom of risk itself imposes a reductive frame. As laypersons from Bristol Bay and beyond come to couch their concerns in scientific registers of risk, these configure and at times constrain what they are able to express. Yet this does not fully explain the more generative forms of political engagement that also accompany the circulation of risk scenarios in Bristol Bay: diverse ways of viewing, valuing, and experiencing the region's ecosystems and social worlds that are present in and even created through the risk-assessment process.

My central contribution lies in understanding how these generative effects arise amid clear constraints, outcomes I locate in the tensions that ensue as everyday experiences fail to conform to the tightly circumscribed parameters that make risk assessment possible. Such tensions exemplify what Michel Callon, Pierre Lascoumes, and Yannick Barthe identify as potentially fecund 'overflows' – unforeseen effects that expose the limitations of existing institutions and frameworks that seek to manage technoscience and what it produces (2011: 9). As the coming sections detail, the Watershed Assessment is built through finely delineated boundaries, which reflect broader impulses of 'boundary-work' separating science from other modes of knowledge (Gieryn 1983; Jasanoff 2005) and scientific from political questions (Jasanoff 2005; Stirling 2005; Wynne 2005). Given the EPA's mandate, the Watershed Assessment establishes narrow frames for public participation and for the spatial, social, and temporal scope of the study – limiting it to the effects of mining on fisheries, for instance. Strained efforts to grapple with these boundaries provoke the very sorts of political contestations that participatory approaches to science-based decision-making are put forth to resolve.[2] I argue that the overflows occasioned by the EPA's risk-assessment exercise have spurred new visions of social, spatial, and temporal connections in turn, suggesting novel possibilities for environmental contests waged in the public domain in Bristol Bay and beyond.

My discussion draws on ethnographic fieldwork conducted primarily from 2012 to 2014 in Bristol Bay communities and in the larger Alaskan city of Anchorage.[3] This research involved semi-structured interviews and participant observation during two key moments: the peer review of a draft version of the EPA's Watershed Assessment in 2012; and the public comment period in 2014 that heard testimony on the EPA's proposal to issue a determination to protect Bristol Bay from large-scale mining. The next section provides a fuller analysis of the literature on processes of risk assessment in participatory forums. I then look more closely at the EPA's Watershed Assessment and how it stages disaster scenarios. In the paper's remaining sections, I move through the social, spatial, and temporal frames that structure the study, analysing salient overflows to the parameters that determine who, what, and when is at issue. In each case, I show how neat boundaries are disrupted by the identities, perspectives, and relations fostered

Journal of the Royal Anthropological Institute (N.S.), 108-126
© Royal Anthropological Institute 2016

by public performances of risk assessment. I conclude that despite the limitations of common formats of consultation and discourses of risk, the negotiation of multiple forms of knowledge and authority in the public view nevertheless opens new spaces and social formations for the exercise of politics.

Public participation in scientific risk assessment

An expanding interdisciplinary literature has accompanied the rapid rise of participatory approaches to decision-making on issues of science and technology, debating their democratizing power and potential (see, e.g., Chilvers 2008; Chilvers & Kearnes 2016; Irwin 2006; Leach, Scoones & Wyne 2005; Stirling 2005; Wynne 2006; 2007). Practices to promote public inclusion first gained currency in the 1970s, but their scope and intensity have grown considerably in recent years, fuelled by both rising public distrust of science and scholarly calls for more democratic decision-making (see Wynne 2007: 100n).[4] Public consultation is now the norm for environmental matters in democracies in the Global North (Chilvers 2008: 2990). In the United States, the EPA has been especially focused on incorporating participatory processes into its risk assessments, particularly in its dealings with Native American tribal governments.[5] The call for 'free, prior, and informed consent' to resource development outlined in the 2007 United Nations Declaration on the Rights of Indigenous Peoples has led to renewed pledges for consultation by extractive industries, including the international mining industry (ICMM 2013). Mechanisms for participation have become so ubiquitous that many of the same scholars who first drew attention to the need for public engagement in scientific decision-making now urge its critical reappraisal (Reno 2011: 845).

Ethnographic studies examining participatory approaches to environmental issues raise a variety of concerns about their potential for offering meaningful inclusion, tracking a broader wave of scepticism about the pursuit of 'participation' more generally (e.g. Cooke & Kothari 2001). Recent work highlights how common instruments for facilitating participation often reinforce existing power relations, depoliticizing struggles by framing problems and procedures in technical terms (see, e.g., Burke & Heynen 2014; Himley 2014; Holifield 2010; F. Li 2009; T. Li 2007). Analysing the EPA's risk-assessment practices, for instance, Ryan Holifield finds that certain consultative mechanisms neutralize more radical efforts to redistribute risk by drawing reductive boundaries around target communities, among other factors (2004: 296). Studies like these underscore how the push for inclusion often has exclusionary effects, as widely implemented strategies for public participation typically rely on boundary-work that maintains technoscientific expertise as the dominant register of legitimate knowledge.

At the same time, ethnographic research also reveals messy and unpredictable outcomes in everyday settings, illustrating how this boundary-work is continually subverted. Studies show how participatory forums are disrupted and delegitimated as frustrated opponents of resource development schemes refuse to engage, staging protests outside public hearings (F. Li 2009); as actors on the margins take up technoscientific tools or roles to pursue their own 'counter'-projects (Hébert & Brock under review; Holifield 2009; Reno 2011); and as plans simply fall flat when unruly subjects fail to heed attempts to remake them into participatory players (Welker 2012). Contestations seem to bubble up amid inclusionary efforts to pin down environmental impacts (see, e.g., Bond 2013; Cepek 2012; Himley 2014; Holifield 2010; Howe 2014;

F. Li 2009; Mathews 2011; Reno 2011; Vaughn 2012). Given the persistent ways in which constraints upon participation are subject to overflows like these, tensions over knowledge and authority are likely to be manifest even within the most tightly organized public forums, and perhaps only more so.

While there has been substantial analysis of participatory approaches to environmental governance, there has been less sustained inquiry into how these intersect with the push for prognosis that marks environmental politics today (see papers by Mathews & Barnes and by Zeiderman, this volume), which is especially apparent in efforts to assess future-orientated risk. Like the recent interest in participation, attention to risk initially grew from efforts to expand the democratic basis for decision-making on issues of science and technology (Wynne 2002). Yet the foregrounding of risk in participatory forums today may contribute to reaffirming the very technocratic underpinnings that consultative initiatives are intended to transform. As Wynne contends, 'extensive, deliberate and well-intentioned participatory moves' can paradoxically obscure the 'subtly anti-democratic implications of translating more multifarious, messy, vernacular and contested issue definitions into monolithic "risk" terms' (2005: 69). Indeed, anthropological research indicates that those labelled as 'at risk' often refuse the framework, preferring to interpret their situation through other rubrics (Cepek 2012; Hébert 2015). As this suggests, conceptualizations of future loss or harm necessarily depend on 'contested issue definitions'. But as risk has grown as a site for intervention, Wynne argues, it has increasingly become treated not as an arena for political struggle but as an objective entity that can be known through scientific investigation, calculated through probabilistic modes of assessment, and controlled through managerial techniques (2002: 467-9).

The problem with this, Wynne and others note, is that it prematurely forecloses the possible futures that might be envisioned through participatory activities. As Callon *et al.* argue, the concept of risk that came to the fore with Ulrich Beck's (1992) influential account of the 'risk society' seems to take for granted that 'researchers and engineers are able to take an inventory of the possible states of the world – in short, to describe the set of likely scenarios' (Callon *et al.* 2011: 227). In Callon *et al.*'s view, this unnecessarily limits politics to 'the choice of a world from those which are known or can be anticipated' and undercuts the possibility that actors might pursue 'a still-unknown common world' by inventing spaces for exchange 'that will enable them to devise scenarios rather than just choose between scenarios' (2011: 227-8). In Wynne's estimation, the drive for public participation often fails to account for the 'material influence' that imagined futures exert on the present (2007: 106), prompting questions about the political effects of the threats conjured through rituals of risk assessment in Bristol Bay.

Wynne's point signals risks of another sort – risks for the practice of politics – embedded in the inclusive approaches to risk assessment currently in vogue. In contrast, Callon *et al.* (2011) emphasize the liberatory prospects of the 'hybrid forums' that comingle expert and lay authority, attributing to them great promise to generate novel forms of knowledge with the potential to destabilize the very category of expertise, such as that guiding risk assessment. While the EPA's efforts in Bristol Bay help reinforce the authority of scientific expertise, the exchanges that take place within and alongside its forums of comingled knowledge do, at moments, express and transform the perspectives of those involved, at times in ways that animate new forms of political engagement. This suggests that the politics of imperilment at work in Bristol Bay, like the politics of endangerment that Timothy Choy theorizes in Hong Kong (2011: 48), can inspire outcomes that exceed the limiting confines of risk discourses. The EPA report is structured around technoscientific scenarios, as outlined in the following section, which set the stage for the overflows to follow.

The EPA's Watershed Assessment and the staging of disaster scenarios

> Uncollected leachates from waste rock piles and TSFs [tailings storage facilities] would elevate instream copper levels and cause direct effects on salmonids [fish of the salmon family] ranging from aversion and avoidance of the contaminated habitat to rapidly induced death of many or all fish ... Rapidly induced death of many or all fish would occur in 12 km (7.4 miles) of streams in the Pebble 6.5 scenario. Copper would cause death or reduced reproduction of aquatic invertebrates in 21, 40 to 62, and 60 to 82 km (13, 25 to 38, and 37 to 51 miles) of streams in the Pebble 0.25, 2.0, and 6.5 scenarios, respectively.
>
> Excerpt from the Watershed Assessment's Executive Summary, from a section that addresses 'risks to salmon and other fishes' from 'water quality: leakage during routine operations' (EPA 2014a: 15)

Through precise, data-driven scenarios, the EPA's Watershed Assessment assembles powerful images of possible future devastation, whose performative force is heightened as they are repeated, elaborated, and disputed in the public forums in which the debate over Pebble unfolds. Insofar as these forecasts turn a teeming abundance of natural resources into a site of acute vulnerability, they invert the usual sense of potentiality embedded in the notion of natural resources. Given the etymology of the English word 'resource' and its general identification as an element to be put to human use, Elizabeth Ferry and Mandana Limbert observe, 'It is as if to define something as a resource is to suspend it between a past "source" and a future "product"' (2008: 6). Yet in the face of risk, or what Beck (1992) famously describes as the growing distribution of 'bads' instead of goods, resource futures are overlaid with the spectre of disaster and projected loss (see also Choy 2011).

Over more than a thousand pages, the Watershed Assessment synthesizes the current science on Bristol Bay, including voluminous baseline data collected by Pebble developers. It provides evidence for 'a largely undisturbed region with outstanding natural, cultural, and mineral resources' (EPA 2014a: ES-3) by way of mostly quantitative metrics: numbers of animal species; figures about annual salmon returns, employment, and industry revenues; statistics involving human populations and demographics; and so on (EPA 2014a: ES-5-ES-9). These data then serve as the basis for its evaluation of the likely consequences of a few different mine scenarios, which vary according to the scale of mining and 'reflect the expected characteristics of mine operations at the Pebble deposit' (EPA 2014a: ES-4). In addition to circulating its draft report among a wide range of experts 'for comment on its technical accuracy and policy implications' (EPA 2012a), the EPA convened a panel of twelve scientists and engineers with varied backgrounds to conduct a formal peer review of the draft document, which took place in conjunction with yet another public hearing held at the peer-review panel meeting itself. The report was revised based on this feedback.

As a study, the Watershed Assessment clearly defines its scope of analysis. It considers several geographical scales, from the Bristol Bay watershed down to the mine footprint, where it examines three different scenarios that vary in the amount of ore processed and years of active mining, labelling them Pebble 0.25, Pebble 2.0, and Pebble 6.5 for the number of billion tons of copper produced over the mine lifetime (EPA 2014a: ES-3-4, 11). It also addresses two time periods: a development and operation phase, estimated to last from twenty to a hundred years or more; and a post-mining phase, which 'could continue for centuries and potentially in perpetuity' (EPA 2014a: ES-4). Further, as the report's title specifies, it does not set out to offer a comprehensive look at the implications of mineral development on all aspects of life in the region, but rather 'an assessment of potential mining impacts on salmon ecosystems of Bristol Bay, Alaska'. The Watershed Assessment makes plain that given the EPA's purview – the federal agency's involvement with prospective mining on state lands is based on its authority under the US Clean Water Act's Section 404(c) – and various feasibility considerations, its scope is limited to the direct effects of mining on fisheries and 'consequent effects on wildlife and Alaska Native cultures in the region' (EPA 2014a: ES-3). It thus considers these other impacts only insofar as they are mediated by fish.

Within the report's limited scope, it specifies potential impacts based on scenarios involving the likely effects of routine mining operations as well as potential failures, such as tailings dam breaches (EPA 2014a: ES-12, 17). In one table, parallel columns organize data to indicate, for example, that the probability of a 'concentrate spill into a wetland' is '2 wetland-contaminating spills in 78 years' with consequences of 'acute exposure to toxic water and chronic exposure to toxic sediment' for invertebrates and, potentially, fish (EPA 2014a: 17). The accompanying text describes these outcomes in pithy summaries that nevertheless build to vivid enactments of future disaster:

> Failure of the [TSF, or tailings storage facility] dam ... would result in the release of a flood of tailings slurry into the North Fork Koktuli River. This flood would scour the valley and deposit many meters of tailings fines in a sediment wedge across the entire valley near the TSF dam, with lesser quantities of fines deposited as far as the North Fork's confluence with the South Fork Koktuli River ... **Near-complete loss of North Fork Koktuli River fish populations downstream of the TSF and additional fish population losses in the mainstream Koktuli, Nushagak, and Mulchatna Rivers would be expected to result from these habitat losses** (EPA 2014a: 21, 22, emphasis in original).

While the precise calculations that ground these scenarios speak to the probabilistic exercise of risk-assessment science, they crescendo into almost cinematic narratives of an imperilled future.[6] On one level, the Watershed Assessment participates in the familiar quantitative depiction of disaster as a 'disembodied scientific object measured against an implemented baseline' (Bond 2013: 709). On another, it presents a departure from this baseline with the kind of dramaturgical flair that recalls the disaster scenarios that Andrew Lakoff terms 'imaginative enactments' (2008: 402). Whereas Lakoff supposes that these are crafted in the absence of possibilities for quantitative risk assessments, such as in the case of flu pandemics, the EPA report shows that imaginative enactments can be woven into the substance of risk assessments themselves. Imaginative enactments may in fact prove a likely feature of such exercises, given that pinpointing risk always entails some form of prognostic work.[7]

Such enactments are not uncontested, of course. In the almost-obligatory battle of 'duelling experts' that characterizes technoscientific disputes today (see Fischer 2005), mining supporters claim that the Watershed Assessment has 'serious methodological and scientific flaws' (PLP 2014). Attacking the EPA's science and the numbers derived from it, they deploy counter-evidence to support alternative projections of techno-utopian futures. Pebble developers assert, for instance, that their research shows that mining development could actually *benefit* fisheries habitat through 'a science-based surplus water release strategy, employing more rigorously devised hydrology estimates and sophisticated PHABSIM [physical habitat simulation] modeling of stream flow-habitat relationships' (PLP 2014).

When mining proponents declare the rigour and sophistication of acronym-packed modelling tools outside the frame of reference of most laypersons, they claim trustworthy, objective, 'science-based' techniques for their own side. The stakes of such claims are high. By invoking findings like those outlined above, Pebble developers successfully lobbied the EPA to launch an internal investigation as to whether the agency inappropriately collaborated with Pebble opponents in developing the Watershed Assessment (Demer 2014). In addition, they have pursued legal action to block the EPA's involvement. Although some of these charges were dismissed, others are still playing out in the courts, essentially freezing the EPA's work in Bristol Bay (Martinson 2015).

As in other environmental contests, battles over the legitimacy of scientific data and the competing stories these support shape the social and environmental terrain that emerges through the Pebble fight. Even more suggestive, however, are the subtler overflows that reveal the limits of existing frameworks for setting the terms of the debate. In the remaining sections, I analyse how efforts to fit Bristol Bay resources and their stakeholders into the neat categories of scientific risk assessments lead to disagreements, discomfort, and other difficulties that speak to the more expansive political possibilities that the Pebble contest has engendered. The following sections detail the parameters and overflows that stem from efforts to bound who, what, and when is at issue.

Who: authoritative spokespersons

[W]hat I have to say in the next few short minutes is truthful, and it comes from my direct experience as a former federal regulator with the US Army Corps of Engineers. I mean, being the only person in this room that's actually processed a Section 404 permit for a mining operation. And please note that I was also a federal 404 enforcement officer working for the Corps and with the EPA for more than ten years. Since my time is so short, I'll very quickly jump right in and explain why I have little respect for EPA and give you a very quick example of how EPA deceives people. Again, please remember that I do not work for Pebble. I'm speaking to you from a position of someone that actually worked and supported EPA – worked for EPA and supported them – or with them as an enforcement officer and an expert on 404 wetland issues.

Comments delivered by Ray Kagel to the EPA at a public hearing on 14 August 2014 in Dillingham, Alaska (EPA 2014d: 82-3)[8]

For over a decade I've witnessed the fight over the Pebble Mine deposit. It's evident that our land is being harassed by outsiders who do not know the area. These intruders do not understand the serious dependency our people have on our local animal populations, especially the salmon.

Comments delivered by Brian Abraham to the EPA at a public hearing on 14 August 2014 in Dillingham, Alaska (EPA 2014d: 44)[9]

The mandate to assemble stakeholders to speak on behalf of Bristol Bay resources provoked revealing complications. Despite the openness of the EPA's participatory proceedings and its well-executed bid for broad-based input – or perhaps because of it – the exchanges that unfolded in public meetings dramatize tensions surrounding authority, expertise, and claims to legitimate knowledge. These overflows highlight the challenges of hewing to the narrow constraints established by the risk-assessment process.

While tensions of authority were in evidence throughout the participatory exercises, as the above quotations from the 2014 hearing in Dillingham suggest, they were especially palpable in the public comment session that was held as part of the 2012 peer review of the Watershed Assessment, where questions of scientific expertise were foregrounded. Like the Dillingham hearing, the peer-review hearing was a highly structured event. Held at a relatively new and upscale convention centre in downtown Anchorage, it required preregistration to attend. A professional facilitator opened the meeting by reminding those present that it was 'all about science and technology'. Registered attendees had received prior 'ground rules' via e-mail that explained that the goal of the session was to 'provide technical and scientific perspectives on the assessment' (EPA 2012*b*). Members of the public were called to the microphone in a predetermined order to deliver their few minutes of comments to the peer reviewers onstage.

By creating a forum to solicit public input on the report's science, the EPA process opened scientific research to participation from those who are not typically involved in rituals like peer review. Because the public testimony was followed by an open discussion among the peer reviewers about the draft report, members of the public had an unusual window onto a process that generally happens, even for scholars, behind closed doors. As the proceedings developed, I became enlisted by Bristol Bay residents I knew as a *de facto* informant on the obscurities of peer-review practices and performances. Those who had come to Anchorage to speak in support of the Watershed Assessment and against Pebble seemed concerned that the report was attracting so much criticism – not simply from members of the public, who included Pebble officials and employees as well as representatives from Alaska's mining lobby and chamber of commerce networks, but more significantly from the peer reviewers themselves. After remarks made by one of the more disgruntled reviewers, the Bristol Bay resident sitting next to me furrowed his brow, and whispered, 'It doesn't look good for us'. I explained to him and a few others seated nearby that a key purpose of peer review is to identify shortcomings in data and analysis and pinpoint opportunities for revision, so it was not surprising that the tone of the discussion was one of criticism rather than praise. The conversation highlighted the uneven familiarity among those in the room with certain expert conventions. Such exchanges, which were held as asides on the margins of the formal process, made clear that the public performance initiated by the hearings opened up scientific knowledge production in new ways but at the same time reinscribed some of its more exclusionary aspects.

This was especially evident in how attendees without scientific training or credentials negotiated the charge to keep their comments 'all about the science'. Most scientists or researchers at the meeting introduced themselves by way of their academic qualifications and professional appointments before speaking on technical domains, from aquatic biogeochemistry to zoology. Those without advanced degrees sometimes acknowledged their lack of such credentials, typically in a somewhat sheepish or apologetic fashion. Others, however, laid claims to knowledge and authority by introducing themselves as 'an expert in my culture and subsistence', for example, or 'an expert in subsistence', referring to the hunting, fishing, and gathering activities that are core to rural and Alaska Native identity. One person offering public comment characterized these lifeways as 'topics I have studied since my birth'. On the one hand, identifying subsistence skills as expertise in the peer-review forum posits them as a source of legitimacy. On the other, the imperative to represent one's authority in a mould of expertise formed with scientific credentials in mind draws attention to the uneven opportunities for participation that were reinforced anew through the peer-review process.

During the public comment session, the mantle of expertise often served to restrain interventions. The powerful groups that were lined up on both sides of the issue, reflected in clusters of pro- and anti-mine adherents seated around the room, appeared to have given their contacts well-honed talking-points in preparation for the event. Yet sticking to these technical scripts proved challenging for certain non-scientists given their different reasons for speaking. For example, during one Bristol Bay resident's remarks to the panel, he broke with the usual talking-points and tenor of the meeting to declare angrily that he had recently spoken with village elders who were 'scared shitless' about the possibility of the mine. Shaking and sputtering, he riveted the room. While opinions varied as to whether his display of emotion was inspiring or inappropriate, it drew attention to the less explicit ground rules that modulated contributions offered from the floor. Although this participant was allowed to complete his comments, the professional facilitator moved in quickly after he had finished to remind the suddenly-more-alert audience of their duty to comment dispassionately on the science. This offers a prime example of what scholars of participatory processes have observed in countless other contexts, which Wynne describes as the 'scientistic diktat of "unpolitics"' that characterizes routinized public meanings (2008: 26).

Despite such efforts to restore tight parameters to the science-based process, the polarized nature of the Pebble issue made their overflow nearly impossible to contain. Tensions about who is authorized to speak for what and for whom, and on what basis, were always close at hand. While the peer reviewers deliberated onstage about whether and how to include certain animal species left out of the draft Assessment, Bristol Bay residents from different parts of the watershed disputed who represented the real 'local knowledge' at issue during their time at the on-floor microphones. Most of the residents living on the west side of Bristol Bay have come out in opposition to the mine, whereas there has been more support for it in certain communities on the east side of the region, which both is closer to the location of the Pebble deposit and has seen more local employment opportunities from exploration and outreach efforts.[10]

Strains within the opposition movement were evident on the sidelines of the proceedings as well. Those who attended the meeting to support the EPA's effort represented the remarkable range of groups that have come together to defeat Pebble, including commercial and recreational fishing interests, environmental non-profits, and Alaska Native organizations – groups that are not always in agreement on much else. While the diversity of these 'unlikely coalitions' has no doubt contributed to their successes (see Snyder 2014: 21-2), it keeps disagreements about who can legitimately speak for Bristol Bay resources not far offstage.

Although the public comments delivered to the peer-review panel by Anchorage-based bird biologists and Alaska Native organization officials may have dovetailed closely with one another, off-microphone conversations at times revealed a less seamless meshing of perspectives. In one encounter at the convention centre, during banter at a recess, a Bristol Bay resident was regaling some friends with tales of extra-legal subsistence fishing under the noses of enforcement officers, a source of ongoing amusement for those who bristle under state resource control. At this point, another attendee who happened to be in earshot, a wetlands scientist from Anchorage, admonished the resident for disobeying fishing regulations. In turn, the resident testily demanded if the interloper knew the salmon statistics for the river in question, informing her that there were no conservation concerns and her rebuke was misplaced. The resident's familiarity with these figures and their strategic deployment as a counter-manoeuvre amplified the sparring and hastened its end. As the scientist backed down, the two irritated parties, nominal partners in the fight to 'save Bristol Bay', left for the break with their respective groups. This awkward exchange dramatizes the contestations that simmered throughout the risk-assessment process over the authority of spokespersons and the identity of the 'ecological resources' (EPA 2014a: ES-6) they claimed to represent.

What: ecological resources at the watershed scale

> What you found in this assessment is a treasure, a national treasure on par with the Grand Canyon, but it is even bigger than that. Because of its uniqueness it's a planetary treasure and if we allow something this wild and this big and this rare to be threatened by something as potentially devastating as this mine, I totally believe that we have lost our way to the kingdom of heaven.
>
> Comments delivered by Chris White to the EPA at a public hearing on 5 June 2012 in Naknek, Alaska (EPA 2012c: 26)

Within the participatory forums of the risk assessment, diverse actors enlisted scientific findings to support accounts of comparative significance that quickly spilled beyond tidy bar graphs of salmon returns and probability tables about tailings dam breaches. Concerned citizens offering testimony invoked EPA data to portray Bristol Bay not simply as a storehouse of regional natural riches but also as a global and transcendent 'treasure', a recurring phrase heard at public hearings. Although the Watershed Assessment sets out to restrict the material scope and the spatial scale of its analysis to the ecological resources of the Bristol Bay watershed (and even more specifically to those of the two major river systems whose headwaters lie in the area of the Pebble deposit) (EPA 2014a: ES-1), it contributes to generating a matter of concern that is 'even bigger than that', as public comments make plain. Its statistics build the Pebble fight into a struggle for salvation through a contest of superlatives: between 'the world's most productive salmon fishery', on the one hand, and 'the threat posed by the Pebble deposit, a mine unprecedented in scope and scale', on the other (EPA 2014b; see also Hébert & Brock under review). This also takes shape as participants in the public process wrestle with how to account for Bristol Bay's diverse life forms within the terms of risk assessment.

The focus of the EPA's study on salmon ecosystems in particular offers a prime example of how risk-assessment frameworks can foreclose fuller pictures of potential impacts, provoking a heightened awareness of these limitations in turn. Long before the EPA began its Watershed Assessment, Bristol Bay's voluminous salmon runs and storied fishing industry made the region synonymous with its salmon. Struggles over mineral development only deepened this, particularly in light of the EPA's authority under the Clean Water Act. Rallying around the region's salmon has proven tremendously mobilizing for broad-based anti-Pebble activism. Yet the apparent unity of this endeavour conceals tensions that surface in efforts to specify what, exactly, is to be protected.

For one, a great many of the public comments heard by the peer-review panel in Anchorage in 2012, as well as in the informal conversations that took place in Bristol Bay in its wake, questioned the way the scope of the EPA study was bounded conceptually. The report's narrow focus on 'fish-mediated risks' received widespread criticism during the public comment period. What about the impacts of mining on other animals, some asked, like the many shorebirds for which Bristol Bay is a site of hemispheric importance?

What about the other kinds of industrial development that would accompany Pebble, from power plants to employee housing? How would all this change the culture of Bristol Bay? The awareness of the interconnectedness of Bristol Bay ecologies made the risk-assessment effort seem all the more urgent to many mining detractors, even as the legitimacy of its predictive rigour appeared to depend on refusing to factor in these very interdependencies.

From the perspective of those who oppose the mine, the Watershed Assessment, if anything, underestimates threats by limiting its analysis to fisheries risks and focusing on impacts from mining alone rather than also considering the associated development activities a large mine would necessitate. No doubt at least in part because of these persistent questions during the peer-review hearings, rationales for the report's highly bounded scope were clarified further in the revisions made between the draft and final documents. For example, an entire paragraph was added to the final report's first few pages to explain its exclusion of certain impacts, such as 'induced development' and 'direct effects of mining on Alaska Natives and wildlife', versus those mediated by fish alone (EPA 2014a: 6).

The geographical scale of the watershed employed by the risk assessment also draws attention to connections and movements that simultaneously unsettle its parameters. In particular, the prominence of hydrological science in the creation of scenarios led Bristol Bay residents, long aware of the way water interlinks the region, to perceive their home in new ways. One mining opponent described how an experience years before of overhearing two researchers discussing early hydrological testing prompted his involvement in anti-Pebble campaigns. According to this resident, reports of nitrate tracer elements not flowing directly downhill, as expected, but 'going everywhere', through 'an interconnected water table', impressed upon him that 'we can't have a disaster ... without it affecting all of Bristol Bay'. This shift in perspective, as he described it, provoked his more certain conviction of possible future disaster: 'At the end of the day, I still think that, if this mining proposal was to go through, there's not a question of *if* it would be devastating to this region and the way we know it to be today. It's just a question of *when*'.

From this new vantage, Bristol Bay appears more expansive and interconnected, as well as more acutely imperilled. This widens the frame and amplifies the import of the Pebble issue to one of planetary and even divine significance for many players involved. The region's resource inventories substantiate its representation as 'one of the last best intact salmon ecosystems on the planet' (EPA 2014d), just as it is home to some of 'the last intact, sustainable, salmon-based cultures in the world' (EPA 2014a: ES-8). Unlike many other 'at risk' sites today, which are often marked by a 'countdown' of endangered life forms resonant of salvage ethnography (Muehlmann 2012: 339), Bristol Bay's vulnerability is indexed by brimming profusion. So pristine as to seem almost prelapsarian, the area emerges from this imagery as a not-yet-ruined world, with a fraught relationship to history and the future as a result. Positioning the region as free from and prior to the devastations of industrial modernity ironically participates in some of the same discourses that construct it as lacking, impoverished, and in want of development – in this case, by mining. As one mine opponent described the campaigns to promote Pebble as a needed source of jobs and revenue: 'It's "let's save the poor, drunk, killing-themselves Natives" ... It's infuriating'. These comments reveal how efforts to save Bristol Bay from one future or another generate fundamental disagreement about what the region is, and what it might become.

Journal of the Royal Anthropological Institute (N.S.), 108-126
© Royal Anthropological Institute 2016

When: from now to perpetuity

> Another thing is the word 'in perpetuity'. I've heard the word a lot and also 'forever'. I don't think humans can comprehend what that means. We throw around 'forever' like it means something. It is like hanging a balloon full of acid over Washington DC that is going to be there forever and some day it might break or pop. You can lose your jobs, lose your livelihood or lose your culture and that word 'forever' is risky and that scares me a bit.
>
> Comments delivered by Colter Barnes to the EPA at a public hearing on 6 June 2012 in Igiugig, Alaska (EPA 2012d: 8, quotation marks added).

The public process adopted by the EPA provoked challenging questions involving *what* and *who*: What resources are to be included for a robust accounting of risk, and who can authoritatively speak for them? But the frame that perhaps generated the most reflection among the public and peer reviewers alike pertained to yet another dimension: *when*. More specifically, the charge to consider the fate of toxic mining waste 'in perpetuity' ushered in questions that pushed at the limits of risk-assessment parameters.

The concept of perpetuity provoked tensions within the text of the EPA report and in the public discussions that surrounded it. Peer reviewers remarked that even the lengthy mining operations phase outlined in the report, loosely bounded from twenty to a hundred years, is far from 'perpetuity'. Indeed, the so-called 'perpetual timeframe' introduced by the Watershed Assessment – a phrase included in the draft version but eliminated from the final report – appears an oxymoron, since it extends the study's temporal horizon into eternity. A *perpetual* timeframe seems not a time *frame* at all, but rather time's un-bounding. Not long after the public was reminded by facilitators to keep comments to technical matters at the peer-review hearing, peer reviewers mused philosophically about how to square existing scientific tools with a consideration of the management of toxic substances in perpetuity, a prospect that one peer reviewer described as 'scary'. 'This is what we all struggle with', he reflected to the group, 'what is *long-term*?' The infinite telescoping-outward of time invoked by perpetuity complicates the already mind-bending temporal scale-jumping demanded by the Watershed Assessment, which combines seismic and geological data that run in the thousands and millions of years with decadal, yearly, and seasonal salmon cycles that make for a potentially indefinitely 'renewable resource'. This is on top of what one person offering public testimony described as the seven-days-a-week, 365-days-a-year activity that makes rural subsistence hunting, fishing, and gathering lifeways 'not a part-time job'.

In the same way that encountering the prospect of Pebble and engaging in the EPA process spurred new visions among the region's residents of spatial and social boundaries and connections, it prompted a reappraisal of temporalities as well. For one, the markedly different rhythms, tempos, and durations – or what Barbara Adam terms 'timescapes' (1998) – bound up in the materialities of particular resources inflect the various possible futures projected in scenarios of risk. Also at play is a shift in the reckoning of time that is introduced by the identification of a mineral prospect, or what Ferry and Limbert, alluding to W.B. Yeats, highlight as the potentiality of 'the stone unquarried' (Ferry & Limbert 2008: 5). Outside the confines of official public hearings, many region residents reflected on changes ushered in by the discovery of the Pebble deposit and the struggle against it, even if mining never happens. As one mine opponent remarked: 'Sadly, because of the resource that is in the ground up there, there will always be pressure. And, you know, who's not to say that there might be some amazing technological development that might allow it to happen in a prudent and safe manner, twenty, thirty, forty years from now?' Another expressed the view, widespread among residents, that 'my projection for the future here is that this Pebble fight is going to be a long protracted fight for generations'. The prospect of mining has transformed senses of time in the region, as many now envision ongoing mobilizations and political battles stretching far into the future.

Bristol Bay residents highlighted in turn how this shifts the burden of looking after the region's future onto younger generations. One younger resident involved in anti-mine campaigns said she hated it when older people told her, 'I'll be dead before this is over', as if they were wishing her 'Good luck with that!' 'Because I don't wanna fight this for the next twenty years', she clarified. Yet some older residents suggested that the entire community was being reorientated toward a longer-term view as a result of the conflict over Pebble. 'We live here, our kids live here', one explained, 'we are instilling values in them as we go through this fight ... So the generation is coming up or being taught that this is a very important fight'. He described how with continued funding and strategic alliances, the struggle against mining proponents might continue 'for years and years'. 'Maybe some time in the future they'll get their way and they'll come in', he continued. 'And hopefully by then they do have the technology not to –', he stopped abruptly, confessing that he did not understand how the mine developers would ever be able to mitigate damage to streams and the aquifer. 'That's the part I really don't understand', he acknowledged, 'but that's another fight for another generation'.

This resident's admission of uncertainty provides another illustration of overflows that are not easily accommodated by formal rituals of risk assessment, whose results are premised on modes of bounding and quantification that tame uncertainties into more calculable probabilities. During the peer-review discussion, panel members expressed discomfort as they thought aloud about the report's mandate to map out likely futures with an air of exactitude. At the same time that they explicitly voiced concern about the implications of large-scale mining for the region, they openly discussed the challenges of evaluating a risk assessment built from multiple extrapolations. As one noted, the report's extensive modelling lends the analysis what he called a 'false sense of precision', given the 'danger' of presenting potential future outcomes 'to a decimal point'. In this respect, the use of scenarios actually winds up calling attention to the sorts of imponderabilities that trouble predictive science, even as they show evidence of mining risks to Bristol Bay.

The issue of uncertainty in fact proved one of the major flashpoints in the peer-review process. Much of the criticism about the Watershed Assessment directed from the floor by mining proponents concerned the draft report's use of 'hypothetical mine scenarios' as the basis for its risk assessment – or what Pebble developers derided as 'speculative mine scenarios' (PLP 2014). In answer to such criticisms, all language of hypotheticals was written out of the final report. This was replaced with repeated reference to 'realistic mine scenarios', with the emphasis that the scenarios 'are based on specific characteristics of the Pebble Deposit and preliminary plans' for mining (EPA 2014a: ES-28). As the revised version of the study explains, the necessary deviation of actual events from the projected scenarios 'is not a source of uncertainty, but rather an inherent aspect of a predictive assessment' (EPA 2014a: ES-28). Yet precisely because of the goal of prediction, the very idea of uncertainty becomes, so to speak, a political minefield. Just lines later, the EPA report notes that the 'large spatial scale and long durations required to mine the Pebble deposit' make a wide variety of 'inherent uncertainties

Journal of the Royal Anthropological Institute (N.S.), 108-126
© Royal Anthropological Institute 2016

more prominent' (2014a: ES-28). This play of knowledge and uncertainty may stem in part from the very charge to assess scientifically the prospect of a 'prospect', whose definition, according to the *American Heritage Dictionary*, slips between 'something expected', and a 'possibility', in its primary definition, and secondarily, in the case of mining, either 'an actual or probable mineral deposit'.

Whereas the play of the possible threatens to snarl the computational forecasts of risk-assessment scenarios, a not-yet-determined future has proven more energizing for Bristol Bay residents. In independent interviews, multiple region residents spoke about how the prospect of Pebble has generated a newfound sense of the importance of, as one put it, 'preserving what we have'. This was described not as maintaining a static resource inventory but as enacting future-orientated aspirations instead: 'We're going to thrive, and things are going to get better', one resident reflected. Such comments about the future were saturated with risk, which weighed heavily on the present: sulphuric acid, heavy metals, and a ruined aquifer, to name but a few of the images Bristol Bay residents associated with the never-ending prospect of mineral development down the line. One of the activists quoted earlier described how 'taxing' it felt to have 'this type of thing hanging over your head constantly, that you're constantly fighting'. But somehow the sense of the future as an outcome still to be determined has led to new energies for political engagement as well, especially among younger residents. 'It's scary but it's also motivating', this activist remarked, ending on a hopeful note. 'I mean, I don't know if it'll be over forever. But I feel like . . . we have too much going here, too many people that are committed to it, to just watch it get pissed away for some gold mine'.

Reflections like these point to the more generative political possibilities that have accompanied the imaginative enactments of the scientific risk assessment, with its expert predictions of potential calamities involving destroyed streams and decimated salmon. They show how Bristol Bay residents have been prompted to reimagine their home, themselves, and their futures, reassembling the environment to be protected along with those who speak in its name. These developments suggest that rituals of public participation staged through the framework of risk do not foreclose the expression and formation of what Wynne calls 'more multifarious, messy, vernacular and contested issue definitions' (2005: 69), even as they shape and constrain them in profound ways.

Conclusion

In this paper, I have analysed how the spatial, social, and temporal boundaries required by prevailing approaches to risk assessment provoke overflows that raise political questions involving vying claims to knowledge and authority. As the Bristol Bay case shows, these tensions emerge through modes of public participation that are characteristic of environmental governance today, marked by the comingling of expert and lay perspectives. As Bristol Bay residents and other members of the public contribute to the EPA's Watershed Assessment, they make visible identities, practices, and relations that complicate its effort to separate scientific and technical truths from political contestations. In addition, the dynamics that ensue prompt them to see the spatial, social, and temporal configurations in which they participate in new ways, signalling generative effects that arise not in spite of but through constraints embedded in the process.

My analysis follows other ethnographic explorations of participation in decision-making on environmental issues in noting that consultative procedures alone appear to do little to change the underlying relations that valorize technoscientific expertise at the expense of alternative claims to knowledge and authority. For this reason, hybrid forums for debating risk in the public view are not likely to shatter existing power imbalances. Yet the exchanges they afford nevertheless seem to allow for greater exploration of the boundaries of the political than do governance rituals that take place more behind closed doors: for example, amid the legal proceedings through which the struggle over Pebble is currently being waged. Evidence from Bristol Bay shows that even though the framework of risk has contained debates over Pebble in the terms of technoscience, it has not foreclosed possibilities for political reimagining. This suggests that despite the restrictions imparted by the discourse of risk, experiences of imperilment can harbour more than political disempowerment.

As I have shown, the fight over the proposed Pebble Mine that has played out in participatory spaces of scientific risk assessment has transformed what many Bristol Bay residents once thought possible. Given that the EPA is poised to potentially block mining in the area around the Pebble deposit, the prospect of keeping powerful transnational mining companies out of this part of southwest Alaska, at least for the time being, suddenly seems conceivable. This outcome is far from certain. But being able to envision it has changed the way the region's residents imagine what the future is likely to hold. Whereas there was a widespread impression in the mid- to late 2000s that it would be hard to defeat Pebble, this view has given way to a future that is more in question. In Bristol Bay today, as Fernando Coronil observes in another context: 'Although the future is not open, it offers openings' (2011: 260).

In the same way that uncertainty about the future has disrupted any sense of inevitability surrounding Pebble's development, it has at least opened the door to the formation of 'a still-unknown common world' (Callon *et al.* 2011: 227). Such openings emerge not from the neatly delineated scenarios and participatory processes that characterize risk-assessment exercises today, but through the connections and relations that exceed those technical parameters, particularly when their exclusion itself becomes a political matter. Bristol Bay provides good metaphors for interpreting the power of such overflows, which ultimately prove as challenging to contain, perhaps, as acid mine drainage in a southwest Alaskan watershed.

NOTES

I am grateful to the organizers and editors of this special issue, and especially to Jessica Barnes, Evan Killick, and two anonymous reviewers for their constructive feedback. I also thank Abigail Neely, Susanne Freidberg, Frank Magilligan, and the other members of the Dartmouth College Department of Geography for their very helpful suggestions on an earlier draft of this paper. Special thanks go to Guntra Aistara, Diana Mincyte, Brian Burke, Karin Friederic, and Andrew deValpine for their careful readings, as well as to Lexi Tuddenham, Taylor Rees, Amy Zhang, and students in my 'Politics of Environmental Knowledge' seminar at Yale's School of Forestry & Environmental Studies in the spring of 2015. This research was conducted with support from the National Science Foundation's Arctic Social Sciences program (Award # 1219390), Yale University's MacMillan Center, and the A. Whitney Griswold Faculty Research Fund. It reflects my ongoing conversations with Danielle DiNovelli-Lang and the rest of the Alaska crew, including Lexi and Taylor, Samara Brock, and Kendall Barbery. Finally, I thank those in Alaska who have participated in this research. All errors and omissions are my own.

[1] These events were attended by thousands of people, all told, and over a million public comments were submitted about the draft Watershed Assessment alone (see EPA 2014c). The EPA's participatory process received a strong positive response from many people in Bristol Bay, even from some who were not wholly supportive of the agency's involvement in the Pebble issue to begin with, making it an especially fitting case for examination.

[2] Similar observations have been made about the role of science in decision-making generally speaking (e.g. Himley 2014; Sarewitz 2004), and of EPA science in particular (Jasanoff 1992).

[3] This study relied on the work of a research team that included K. Alexandra Tuddenham, Kendall Barbery, Samara Brock, and Austin Lord to conduct more than ninety semi-structured interviews in the summers of 2013 and 2014. This paper is also informed by the long-term ethnographic fieldwork I have done in Bristol Bay since 2003.

[4] Burke and Heynen (2014) also provide an overview of these trends. Alaska was a prominent site for citizen-science engagement in the early 1970s, as this was a noteworthy feature of debates over the construction of the Trans-Alaska oil pipeline (Coates 1993).

[5] Work by Sheila Jasanoff (1992) and Ryan Holifield (2004; 2010; 2012) explores how EPA approaches to risk assessment have emerged in response to prior challenges and critiques, embedding multiple motivations in current practices and keeping them subject to ongoing contestation.

[6] See Hébert and Brock (under review) for further discussion of the relationship of numbers and narrative in the Pebble controversy.

[7] I thank Brian Burke for raising this point.

[8] Kagel describes himself as 'a certified professional wetlands scientist and a fish and wildlife biologist' who with his wife runs an environmental consulting company based in Idaho. He tells the audience he has come to testify 'at the invitation of our Native American friends at Iliamna', a village near the Pebble deposit known to support the mine (EPA 2014*d*: 82).

[9] Abraham describes himself as a resident of the Bristol Bay community of Togiak and the EPA's Indian Governmental General Assistance Program (IGAP) co-ordinator there.

[10] Survey data indicate that 70 to 80 per cent of Bristol Bay residents oppose the proposed Pebble Mine (Snyder 2014: 22).

REFERENCES

ADAM, B. 1998. *Timescapes of modernity: the environment and invisible hazards.* London: Routledge.

BECK, U. 1992. *Risk society: towards a new modernity* (trans. M. Ritter). London: Sage.

BOND, D. 2013. Governing disaster: the political life of the environment during the BP oil spill. *Cultural Anthropology* **28**, 694-715.

BURKE, B.J. & N. HEYNEN 2014. Transforming participatory science into socioecological praxis: valuing marginalized environmental knowledges in the face of the neoliberalization of nature and science. *Environment and Society: Advances in Research* **5**, 7-27.

CALLON, M., P. LASCOUMES & Y. BARTHE 2011. *Acting in an uncertain world: an essay on technical democracy.* Cambridge, Mass.: MIT Press.

CEPEK, M. 2012. The loss of oil: constituting disaster in Amazonian Ecuador. *Journal of Latin American and Caribbean Anthropology* **17**, 393-412.

CHILVERS, J. 2008. Environmental risk, uncertainty and participation: mapping an emergent epistemic community. *Environment and Planning A* **40**, 2990-3008.

——— & M. KEARNES (eds) 2016. *Remaking participation: science, environment and emergent publics.* New York: Routledge.

CHOY, T. 2011. *Ecologies of comparison: an ethnography of endangerment in Hong Kong.* Durham, N.C.: Duke University Press.

COATES, P. 1993. *Trans-Alaskan pipeline controversy: technology, conservation, and the frontier.* Fairbanks: University of Alaska Press.

COOKE, B. & U. KOTHARI (eds) 2001. *Participation: the new tyranny?* London: Zed Books.

CORONIL, F. 2011. The future in question: history and utopia in Latin America (1989-2010). In *Business as usual: the roots of the global financial meltdown* (eds) C. Calhoun & G. Derluguian, 231-92. New York: University Press.

DEMER, L. 2014. EPA watchdog reviewing agency's work on Bristol Bay and Pebble mine. *Alaska Dispatch News*, 6 May (available on-line: *http://www.adn.com/article/20140506/epa-watchdog-reviewing-agency-work-bristol-bay-and-pebble-mine*, accessed 8 January 2016).

EPA [US ENVIRONMENTAL PROTECTION AGENCY] 2012*a*. An assessment of potential mining impacts on salmon ecosystems of Bristol Bay, Alaska, Vol. 1 – Main Report. Seattle, Washington: US EPA. First External Review Draft, May (available on-line: *http://cfpub.epa.gov/ncea/bristolbay/recordisplay.cfm?deid=241743*, accessed 17 January 2016).

——— 2012*b*. External peer review meeting for an assessment of potential mining impacts on salmon ecosystems of Bristol Bay, Alaska: meeting overview, goals, and expectations. Document e-mailed to author on 2 August by Versar Inc. entitled 'Ground rules Bristol Bay meeting'.

———— 2012c. US Environmental Protection Agency, Draft Bristol Bay Watershed Assessment, record of Public Comment Meeting, Naknek, Alaska, Tuesday, 5 June.

———— 2012d. US Environmental Protection Agency, Draft Bristol Bay Watershed Assessment, record of Public Comment Meeting, Iguigig, Alaska, Tuesday, 6 June.

———— 2014a. An assessment of potential mining impacts on salmon ecosystems of Bristol Bay, Alaska. Seattle: Region 10 Office (available on-line: *http://cfpub.epa.gov/ncea/bristolbay/recordisplay.cfm?deid=253500*, accessed 8 January 2016).

———— 2014b. EPA moves to protect Bristol Bay fishery from Pebble Mine (available on-line: *http://yosemite.epa.gov/opa/admpress.nsf/names/r10_2014-2-28_bristol_bay*, accessed 8 January 2016).

———— 2014c. Public involvement for Bristol Bay 404c process (available on-line: *http://www2.epa.gov/bristolbay/public-involvement*, accessed 8 January 2016).

———— 2014d. Public hearing on proposed determination of the US Environmental Protection Agency Region 10 pursuant to Section 404(c) of the Clean Water Act Pebble Deposit Area, Southwest Alaska, 14 August, commencing at 5:15 p.m., Volume I, Dillingham High School/Middle School Gymnasium.

FERRY, E.E. & M.E. LIMBERT 2008. Introduction. In *Timely assets: the politics of resources and their temporalities* (eds) E.E. Ferry & M.E. Limbert, 3-24. Santa Fe, N.M.: School for Advanced Research Press.

FISCHER, F. 2005. Are scientists irrational? Risk assessment in practical reason. In *Science and citizens: globalization and the challenge of engagement* (eds) M. Leach, I. Scoones & B. Wynne, 54-65. London: Zed Books.

GIERYN, T. 1983. Boundary-work and the demarcation of science from non-science: strains and interests in professional ideologies of scientists. *American Sociological Review* 48, 781-95.

HÉBERT, K. 2015. Enduring capitalism: instability, precariousness, and cycles of change in an Alaskan salmon fishery. *American Anthropologist* 117, 32-46.

———— & S. BROCK under review. Mobilizing maps and counter-maps: resource representations, quantification, and the raw material of environmental claimsmaking.

HIMLEY, M. 2014. Monitoring the impacts of extraction: science and participation in the governance of mining in Peru. *Environment and Planning A* 46, 1069-87.

HOLIFIELD, R. 2004. Neoliberalism and environmental justice in the United States Environmental Protection Agency: translating policy into managerial practice in hazardous waste remediation. *Geoforum* 35, 285-97.

———— 2009. How to speak for aquifers and people at the same time: environmental justice and counter-network formation at a hazardous waste site. *Geoforum* 40, 363-72.

———— 2010. Regulatory science and risk assessment in Indian country: taking tribal publics into account. In *Geographies of science*, vol. 3, *Knowledge and space* (eds) P. Meusburger, D. Livingstone & H. Jöns, 231-46. Dordrecht: Springer Science & Business Media.

———— 2012. Environmental justice as recognition and participation in risk assessment: negotiating and translating health risk at a Superfund site in Indian country. *Annals of the Association of American Geographers* 102, 591-613.

HOWE, C. 2014. Anthropocenic ecoauthority: the winds of Oaxaca. *Anthropological Quarterly* 87, 381-404.

ICMM [INTERNATIONAL COUNCIL ON MINING AND METALS] 2013. Indigenous peoples and mining: position statement (available on-line: *http://www.icmm.com/document/5433*, accessed 8 January 2016).

IRWIN, A. 2006. The politics of talk: coming to terms with the 'new' scientific governance. *Social Studies of Science* 36, 299-320.

JASANOFF, S. 1992. Science, politics, and the renegotiation of expertise at EPA. *Osiris* 7, 195-217.

———— 2005. *Designs on nature: science and democracy in Europe and the United States.* Princeton: University Press.

LAKOFF, A. 2008. The generic biothreat, or, how we became unprepared. *Cultural Anthropology* 23, 399-428.

LEACH, M., I. SCOONES & B. WYNNE (eds) 2005. *Science and citizens: globalization and the challenge of engagement.* London: Zed Books.

LI, F. 2009. Documenting accountability: environmental impact assessment in a Peruvian mining project. *PoLAR: Political and Legal Anthropology Review* 32, 218-36.

LI, T. 2007. *The will to improve: governmentality, development, and the practice of politics.* Durham, N.C.: Duke University Press.

MARTINSON, E. 2015. With new ruling in US District Court, Pebble Mine case advances. *Alaska Dispatch News*, 4 June 4 (available on-line: *http://www.adn.com/article/20150604/new-ruling-us-district-court-pebble-mine-case-advances*, accessed 8 January 2016).

MATHEWS, A.S. 2011. *Instituting nature: authority, expertise, and power in Mexican forests.* Cambridge, Mass.: MIT Press.

MUEHLMANN, S. 2012. Rhizomes and other uncountables: the malaise of enumeration in Mexico's Colorado River Delta. *American Ethnologist* **39**, 339-53.

PLP [PEBBLE LIMITED PARTNERSHIP] 2014. EPA's pre-emptive overreach on Pebble is premature and undermines US environmental permitting system (available on-line: *http://corporate.pebblepartnership. com/perch/resources/plp-epa-submission-backgrounder-apr-2014-3.pdf*, accessed 8 January 2016).

RENO, J. 2011. Managing the experience of evidence: England's experimental waste technologies and their immodest witnesses. *Science, Technology & Human Values* **36**, 842-63.

SAREWITZ, D. 2004. How science makes environmental controversies worse. *Environmental Science and Policy* **7**, 385-403.

SNYDER, S. 2014. Bristol Bay wild salmon, Pebble Mine, and intractable conflict: lessons for environmental studies and sciences. *Environment* **56**:2, 17-26.

STIRLING, A. 2005. Opening up or closing down? Analysis, participation and power in the social appraisal of technology. In *Science and citizens: globalization and the challenge of engagement* (eds) M. Leach, I. Scoones & B. Wynne, 218-31. London: Zed Books.

VAUGHN, S. 2012. Reconstructing the citizen: disaster, citizenship, and expertise in racial Guyana. *Critique of Anthropology* **32**, 359-87.

WELKER, M. 2012. The Green Revolution's ghost: unruly subjects of participatory development in rural Indonesia. *American Ethnologist* **39**, 389-406.

WYNNE, B. 2002. Risk and environment as legitimatory discourses of technology: reflexivity inside out? *Current Sociology* **50**, 459-77.

——— 2005. Risk as globalising 'democratic' discourse: framing subjects and citizens. In *Science and citizens: globalization and the challenge of engagement* (eds) M. Leach, I. Scoones & B. Wynne, 66-82. London: Zed Books.

——— 2006. Public engagement as means of restoring trust in science? Hitting the notes, but missing the music. *Community Genetics* **9**, 211-20.

——— 2007. Public participation in science and technology: performing and obscuring a political-conceptual category mistake. *East Asian Science, Technology and Society: an International Journal* **1**, 99-110.

——— 2008. Elephants in the rooms where publics encounter 'science'? A response to Darrin Durant, 'Accounting for expertise: Wynne and the autonomy of the lay public'. *Public Understanding of Science* **17**, 21-33.

Chronique d'une catastrophe annoncée : évaluation du risque scientifique, participation du public et politique de la mise en danger à Bristol Bay en Alaska

Résumé

Comme beaucoup de controverses écologiques du moment, celle qui entoure le projet d'exploitation minière de Pebble Mine, dans une région d'Alaska productrice de saumon, est centrée sur le développement et la contestation des projections scientifiques du risque. L'article retrace la procédure d'enquête publique entourant l'évaluation du risque d'impact environnemental de l'exploitation minière afin d'examiner la manière dont les forums réunissant paroles d'experts et connaissances des profanes donnent forme à des scénarios de périls futurs et influencent la politique environnementale au présent. L'auteure s'appuie sur une recherche ethnographique pour analyser comment les évaluations des risques nécessitent une délimitation des frontières spatiales, sociales et temporelles à l'origine de tensions ou de « débordements » qui révèlent les contraintes des cadres existants. Dans le débat à propos de Pebble, la procédure publique a produit des débordements qui mettent en évidence les conflits entre revendications du savoir et de l'autorité. Elle reflète ainsi le vaste (et souvent vain) travail mené dans le cadre de l'évaluation des risques pour séparer les vérités scientifiques et techniques des contestations politiques. L'article montre comment ces débordements ont eu des effets génératifs, suscitant de nouvelles visions qui reconstruisent les relations spatiales, sociales et temporelles face au danger. Il avance que malgré les limites des procédures de consultation courantes et des discours sur le risque, la négociation de formes multiples de savoir et d'autorité dans l'opinion publique peut ouvrir de nouveaux espaces et de nouvelles formations sociales pour le jeu politique.

Journal of the Royal Anthropological Institute (N.S.), 108-126
© Royal Anthropological Institute 2016

7

A doubtful hope: resource affect in a future oil economy

GISA WESZKALNYS *London School of Economics and Political Science*

In global debates about natural resource extraction, affect has played an increasingly prominent, if somewhat nameless, role. This paper proposes a theorization of resource affect both as an intrinsic element of capitalist dynamics and as an object problematized by corporate, government, and third-sector practice. Drawing on ethnographic research in São Tomé and Príncipe (STP), I explore the affective horizons generated by the prospect of hydrocarbon exploration: a doubtful hope comprised of visions of material betterment, personal and collective transformation, as well as anticipations of failure, friction, and discontent. I also examine the multitude of oil-related campaigns, activities, and programmes initiated by non-governmental organizations and global governance institutions in STP, animated by the specific conundrums presented by oil's futurity. In light of this, I argue that what we see emerging is a new resource politics that revolves around not simply the democratic and technical aspects of resource exploitation but increasingly their associated affective dissonances and inconsistencies.

How do people sense prospective oil? What do they expect from it? And how does this relate to the indeterminacies that characterize contemporary petroleum production? These questions link conceptions of futurity – that is, imaginations of how we might be and associated ethical orientations towards likely and unlikely things to come – with petroleum source rocks, exploration technologies, and calculations of commercial viability, and with concepts of wealth, labour, and redistribution in extractive economies. Importantly, these questions cast oil not as an abstract commodity form but as entangled in the affective fabric of contemporary economic life.[1] This paper examines how lithospheric dreams[2] of wealth, progress, and modernity emerge alongside anxieties and doubts about uncertain outcomes. In doing so, it also reveals a mounting preoccupation borne out by the accounts and practices of public, corporate, and third-sector agencies with the potential for an excess of affect, specifically, in the context of hydrocarbon exploration.

Attention to the affective resonances of natural resources, I argue, has to be part of an effort to comprehend resource environments as simultaneously social, aesthetic, and deeply politicized as well as materially and biophysically grounded (cf. Richardson &

Weszkalnys 2014; West 2005: 639). It expands on a relational understanding of human and nonhuman materialities as mutually imbricated and emergent. Elizabeth Ferry and Mandana Limbert have highlighted the 'affective qualities and moral sentiments' (2008: 12) embedded in the material practices of resource making. They include the contrasting moral sensibilities of a Lockean aesthetic of resource use, nostalgic longings for pristine nature, as well as nationalist projects that deploy natural resources in cosmogonic narratives and visions of a collective future. In Ferry and Limbert's words, 'resources make time material' (2008: 15). More generally, a growing number of economic anthropologists now attend to the ways affect has been implicated in the knowledge practices and productive processes of past and present capitalist societies (e.g. Richard & Rudnyckyj 2009; Tsing 2005; Yanagisako 2012; Zaloom 2009). Affect in this work is seen not as an anti-economic response but as one force among many that give economies their specific shape, while also being shaped by them. Expanding on this work, the analytical questions I wish to probe could be posed as follows: to what extent is it useful to examine the history of natural resource production under contemporary capitalism not just in geopolitical terms but in terms of the forms of affect generated by it? Specifically, what light does this shed on the diverse ethical orientations implicated in imaginations of particular resource futures?

I show here that resource affect may take a variety of forms: euphoria, excitement, aggression, doubt, trepidation, frustration, disillusionment, and so on. These regularly emerge, successively or alongside each other, in contexts of resource prospecting and extraction. Images of the Klondike stampede come to mind, contemporary African mining rushes, the North Dakota fracking boom, or tales of gold-diggers and illicit loggers in the Indonesian rain forest. What is generated, there, may be described as an atmosphere charged with affective force – individually sensed, as Brian Massumi (2002: 28) puts it, but not individually owned. Affect is registered but, in contrast to emotions, not necessarily in a discursive way (Mazzarella 2009: 292). The kind of affects discussed here emerge from the unfolding of specific resource materialities (Richardson & Weszkalnys 2014). Those resource materialities encompass the diverse epistemological, corporeal, and practical modes by which resources are made, the social relationships they reveal, and the generativist ontologies that underpin them.

In some sense, 'affect attaches all over the place' (Hemmings 2006: 559). It is, then, the ethnographer's task to chart the specificity of this attachment in its different cultural and historical inflections. Against affect's dominant theoretical connotations as autonomous, extra-social, and positively transformative force, my account registers its emergence at contingent intersections of corporate, governmental, and social processes and in the midst of existing relations of power (see also Anderson 2010; Navaro-Yashin 2012: 19). Although often treated as mere externality to economic activity, affective states such as hope or disillusionment are better seen as internal to them. In addition, they are subject to both individual and collective moral judgement, visible in the persistent efforts to contain, counteract, and suppress them (Hemmings 2006: 564). I argue that the apparently excessive affective responses generated by resource extraction have progressively entered into a more global consciousness that informs corporate policy and state practice. This is most obvious, perhaps, in instances where community protest against extraction is being channelled into purportedly more manageable forms by juro-political techniques, such as community consultation and compensation payments, in order to safeguard continuous capital accumulation. Yet, as I show later, even where natural resource exploitation is only anticipated, contests proliferate over the appropriate ethical orientations towards extractive futures.

I examine this process, for the most part, through the example of São Tomé and Príncipe (STP), a micro-island state situated in the oil-rich Gulf of Guinea. In STP, the affective capacities of natural resources – especially oil – have been palpable. Contracts for deepwater offshore exploration were signed in the late 1990s. STP was poised to become an oil-producer state within a few years, accompanied by a wave of euphoria about the prospect of imminent wealth and development. In many ways, STP constitutes an archetypal nature-exporting society (Coronil 1997: 7), struggling to come to terms with its colonial legacy of primary-commodity production for a global market. Discovered by the Portuguese in the late fifteenth century, the islands soon became colonial prototypes for the agro-economic exploitation of tropical nature. Their rich soils, climate, and ready access to slave labour markets on the African mainland made them ideal hosts for the first successful plantation economies centred on sugar cane and, later, coffee and cocoa. The cocoa boom of the late 1800s propelled the tiny Portuguese colony into the limelight of global markets. The Creole population, which was consolidated in periods of colonial abandonment and which, in its relation to land and sea, its genetic make-up and sociality, embodied all the contradictions of enforced mixed-race relations and manumission in the Portuguese empire, found itself dispossessed by white Portuguese landowners backed by metropolitan and international finance. Supported by tens of thousands of *serviçais*, so-called 'contract labourers', the plantation economy continued until the end of the colonial period in 1975 (Clarence-Smith 2000).

Following political independence, STP's plantations were initially brought under state control, but later privatized in the context of far-reaching economic reforms with few lasting positive effects. At the start of the new millennium, the country hovers among the world's smallest economies,[3] lacking economic diversification, and is largely dependent on resource revenues, ground rents, and financing by bilateral donors and multilateral organizations (for details, see Seibert 2015). Poverty levels remain well above 60 per cent. Santomeans feel the pressure of their diminishing buying power, exacerbated by their country's insularity, its dependence on imports, and the absence of an economy of scale. With its politics marred by near-constant power struggles and two failed *coups d'état* in 1995 and 2003 (Seibert 2006), Santomeans, though proud of their peaceful democratic tradition, are also dissatisfied with the lack of concrete results. In this context, lamentations about vote-buying, delinquency, and moral decline readily trail off into nostalgic longings for luso-colonial order or for post-independence socialist discipline and respect. They echo familiar postcolonial anxieties over the effects of nominally free politics and markets whose development has not taken the wished-for turns (Mbembe & Roitman 1995: 339).

The hope for oil has not been unequivocal. Comparing STP to the nearby oil producer Equatorial Guinea, the political scientist Ricardo Soares de Oliveira writes: 'These are truly bottom-of-the-pile states with a razor-thin empirical existence. Their mainly agricultural economies were already in a shambles on the eve of oil investment and their bureaucracies are corrupt, unpaid and untrained' (2007: 222-3). In both cases, Soares de Oliveira asserts, companies had little problem negotiating suitable deals with local governments – their investments backed by international risk financing. While the strategy has proven very successful in Equatorial Guinea, oil exploration in STP has languished in an extended pause (Weszkalnys 2015). None of the exploration wells

Journal of the Royal Anthropological Institute (N.S.), 127-146
© Royal Anthropological Institute 2016

drilled between 2006 and 2012 has led to a commercial discovery; as a result, corporate interest in the island state has been muted.

STP's situation is dissimilar from those usually studied by anthropologists of oil, where topics such as community dispossession, environmental destruction, and political strife dominate. Methodologically, I often experienced my fieldwork as an archaeology of affect[4] in the anticipation of oil. It worked backwards from concerns and activities I encountered in STP to reveal the affects that compelled them.[5] The concerns that arise in such prospective situations bear some semblance, imaginatively and practically, to those that have framed extraction in established oil-producer states (Behrends 2008; Weszkalnys 2011; 2014). But they are also centrally about the specific ethical and affective conundrums presented by oil's futurity. For example, the campaigns initiated by non-governmental organizations (NGOs) and other global agencies, such as the Extractive Industries Transparency Initiative (EITI), reflected not only worry about the possibilities for corruption and civil society's capacity to deal with the uncertain future engendered by oil, but also an effort to keep destabilizing euphoria and speculation in check. Millions of dollars have been invested as part of a World Bank governance and capacity-building programme in STP to design institutional infrastructures and a state-of-the-art oil revenue management law. Civil society activists I met suggested that, *oxalá*, God willing, these mechanisms would achieve the desired aims; however, if they failed, the same activists knew that oil would bring only misery, given its tendency, now much discussed in international policy circles, to incite a so-called resource curse comprised of corruption, greed, and social unrest (Weszkalnys 2011). In 2007, not just one but two studies – one initiated by the International Monetary Fund (IMF), the other by the World Bank – scrutinized the potential for 'local content' in STP's nascent oil sector (Klueh, Pastor, Segura & Zarate 2007; MundiServiços n.d.). Reflecting on the presentation of the results in a seminar held in April 2007, an information bulletin issued by STP's National Oil Agency concluded: 'What is important now is to act quickly … because petroleum exploration does not sympathize with the inertia and disorientation that characterizes life in São Tomé and Príncipe'.

The remainder of this paper tracks the surging and dissipating affect of Santomean oil across several years of observations, conversations, institutional reports, news items, and email exchanges. It sketches the contours of the specific historical juncture that allowed an enthusiastic response to oil to emerge, but also what led to its quick transformation into a doubtful hope. My purpose is to query what the Santomean case might tell us about the ways resource affect has become defined, constrained, and channelled in specific directions by a proliferating set of policy-orientated discourses and practices concerned with the political, economic, and societal effects of resource wealth. I analyse notions of expectation management and 'local content' to draw attention to a central issue: that of non-productive wealth, and how it has been seen to jeopardize the unfolding of a prosperous resource future. First, however, I want to outline briefly what analytic purchase a concept of resource affect may have, drawing on existing accounts of resource extraction.

Resource affect

In Guiana … all the rocks, mountains, all stones in the plains, woods, and by the rivers' sides, are in effect thorough-shining, and appear marvellous rich; which … are the true signs of rich minerals,

... no other than *El madre del oro*, as the Spaniards term them, which is the mother of gold, or, as it is said by others, the scum of gold.

Sir Walter Raleigh,
The discovery of the large, rich, beautiful empire of Guiana, 1596

Petroleum ... is a disease that has infiltrated the [Santomean] population, a psychological disease. It's a psychological disease, because people used not to worry, they weren't thinking about anything. Now, with this petroleum, people became sick ... and with this disease, São Tomé and Príncipe is never going to move forward.

Santomean informant, November 2007

From the myth of El Dorado, fuelled by the seemingly insatiable desires of European explorers, to the disease of oil that has afflicted Santomeans, resource affect has carried many names throughout the history of capitalist resource exploitation. This section invites a more careful look at the specific affective relations between the agents and witnesses of resource extraction and its objects – that is, the resource substances themselves – which have figured prominently in both popular and scholarly accounts. I pull together a range of work that has recorded exhilaration and hope, anxiety and disillusionment about the prospect of resource wealth and the future it portends, but that has rarely analysed such affective responses in a comparative or systematic fashion.

The gold rushes of the nineteenth and twentieth centuries epitomize the affective power of precious minerals, triggering extravagant dreams of fortunes to be made and impulsive reactions that were easily contagious. 'It was the age of discovery', notes one account of the Australian gold rush of the 1850s. 'Adventure gripped a whole people like an epidemic. Greed and violence marched with courage and fortitude. Men defied hardship and danger' (Paull 1963: 1). But not everyone was affected alike. The frantic activity of the rush offered opportunity to the penniless, unskilled, and disenfranchised, who had few perspectives otherwise. Neither miners nor speculators, these were people in search of a livelihood. Hundreds of thousands of men, but also women, were drawn to the sites of discovery, arriving on 'ships from London, Liverpool, Glasgow, Hamburg, Marseille, San Francisco and Canton' (Paull 1963: 1-2) in the hope to claim their share of the fabulous riches. Yet the uncertainty of their fortune was palpable, and the threat of forfeiting one's newly acquired wealth constant:

> The prospector, being a perpetual optimist, pursued elusive fortune from one rush to the next, amassing a few hard-won ounces of gold dust in one place to dissipate or lose it elsewhere. His was a precarious existence that paid or withheld rewards for patience and endurance with equal fickleness (Paull 1963: 5).

Rushes and fevers enthral with auspicious, even liberating, force, but they are also viewed with deep ambivalence. Huge effort and hardship in return for what often remains little more than a fantasy can seem a foolish undertaking. A 'pact with the devil' might be able to satisfy immediate desires but ultimately remains barren (Nash 1993 [1979]; Taussig 1980). Money earned from extraction easily unsettles modes of sociality and systems of exchange, leading to the inevitable exclusion of some from established circuits of wealth (Gilberthorpe, Clarke & Sillitoe 2014; Jorgensen 2006; Walsh 2003). Boomtowns are places where, as Malagasy sapphire miners put it, people 'meet when grown' and where trust is, therefore, in short supply (Walsh 2012: 239). The rush, as Andrew Walsh (2003) suggests, creates an atmosphere saturated with heightened speculation but also a sense of shared purpose, expressed in specific modes of risking and daring, which speak of the miners' attunement to the uncertainty of the moment and, not least, to the capricious *hiagna* (will, agency) of the sapphires themselves. However, fantastic stories of El Dorado, bonanzas, and *fofocas* are also sources of suspicion and deceit. Neither one's collaborators nor one's friends and spouses may be reliable companions (Cleary 1990; Paull 1963; Tsing 2005; Walsh 2003).

Affective dissonance of this kind has always accompanied resource exploitation. It is the counterpoint to the sort of affirmative affects that ground nation-building projects in the timeless past of natural resource assets or in a modern pursuit of resource exploitation as a foundation for future progress (Coronil 1997; Ferry & Limbert 2008; Shever 2012). First, visions of an affluent future may throw into relief the ubiquitous signs of 'abandonment and general decomposition' (cf. Mbembe & Roitman 1995: 328), and things that are found wanting. A friend's exclamation that, one day, STP will be a paradise in Africa compels daydreaming because it highlights what, today, are blatant absences: insufficient health services, erratic infrastructures, and understaffed, ill-equipped schools. Similarly, in the face of warnings of an impending 'resource curse' and 'Dutch disease', my Santomean research companions would often declare hydrocarbon production a trump card able to galvanize more sustainable and long-term development opportunities, including fisheries, agriculture, and tourism.[6] What a contrast it would be to the bananas that are now not being planted, to overflowing mango trees, and to the sweet, putrid smell of unpicked fruit lying on the ground! John Locke seems to be speaking from between the lines of this Santomean commentary, seeing the wealth of nature put to its proper use by means of human labour (Ferry & Limbert 2008: 13). Embedded here are both developmentalist dreams built on the transformation of existing resources into prosperity and a moral self-critique highlighting current lack, inadequacy, and waste.

In addition, dissonance arises where resources' generative capacities are marred by experiences of their destructive force. Scarred mountaintops, blow-outs, sacred sites invaded by outsiders, and toxic tailings that slosh where fish used to provide sustenance for riverine people raise questions about both the ecological and cosmological impact of extraction (Gardner 2012; Kirsch 2006; Walsh 2003; West 2006). Such transgressions cause terror and disquiet as a breach both of interhuman reciprocity and of humans' relationship with the environment and important other-than-humans that dwell in it, and are noticed not just by those who fail to share the tales of unimaginable riches or to benefit directly from them (High 2013).

One perennial problem has been how to deal with the apparent excesses of affect and their often confusing and destructive consequences. People have devised a variety of techniques for negotiating and rearranging their destabilized worlds, both by taking recourse to time-honoured rituals of appeasement and by inventing new ones (Bryceson, Jonsson & Sherrington 2010; High 2013; Nash 1993 [1979]; Walsh 2003). We might even reread corporate social responsibility strategies and compensation payments in this light. For the Yonggom of Papua New Guinea, these imperfect attempts to pacify angered locals and to offset ecological (and cosmological) damage and loss of livelihood caused by copper mining fit into a moral economy that likens mining to sorcery. Mining, too, is a practice associated with unrequited responsibility and harmful, often deadly, effects (Kirsch 2006: 108, 208). Compensation claims, in turn, take on the role of earlier cargo cults: a dissonant interface of Western capitalist and indigenous exchange practice. As a result, resources seem to attract riches without any labour input and to result in *moni rain* (a term borrowed from popular pyramid schemes), consumer goods, cars, houses, and a Euro-American lifestyle. Yet, even though *moni rain* may offer temporary relief, it rarely provides a solution to the problem of designing long-term futures (see also Gilberthorpe *et al.* 2014).

Shamanistic rituals, compensation payments, as well as the self-conscious forms of expectation management I shall discuss below, all contribute to the constitution of resource affect as an object of ethical, political, and scientific thought and action. Clearly, what I want to refer to here as the problematization of resource affect has taken different forms in particular times and places (Foucault 2000). A preoccupation with affect's possible excesses, more broadly, and appropriate ways to manage them is not entirely new to the social sciences, but can be traced back to the writings of Gabriel Tarde, Elias Cannetti, and other scholars (Blackman 2012; Mazzarella 2009). They wrote about how to measure, retard, adjust, or govern affective capacities. While Gustave Le Bon experimented with the use of hypnosis, Gabriel Tarde pondered the role of the figure of the leader in facilitating processes of imitation, 'which would hinder or thwart the proliferation of socialism, revolution and terrorism' (Blackman 2012: 35). I return to this issue later, arguing that popular affective responses to resource exploration and extraction have increasingly come into the focus of a loosely aligned set of corporate, state, and third-sector agencies. I have found my Santomean research pulled into a rapidly growing problem field that seeks to gauge and contain potential disruptions, sometimes before the resource is even flowing. First, however, I want to examine more closely the processes that generated dissonant resource affects in STP following the announcement of oil.

Oil's fading promise

When I started fieldwork in STP in 2007, the country's oil prospects were beginning to crumble. In the 1990s and early 2000s, the government had signed various exploration agreements with large multinational (ExxonMobil, Chevron), mid-sized independent (Addax, Anadarko), and smaller, partly speculative companies. An exclusive economic zone (EEZ) had been delineated as well as a maritime development zone operated jointly with Nigeria (the so-called 'joint development zone' or JDZ). However, several exploration wells drilled by Chevron, Addax, Sinopec, and Total in 2006, 2009, 2010, and 2012 only confirmed a lack of commercial viability. Poised between the 'no longer' of a postcolonial plantation economy and the 'not yet' of oil, different horizons of expectation have emerged, partly generated by the apparent failure of a range of scientific and political techniques to reliably portend the country's oil future. Events such as the 14th United Nations Conference on Trade and Development (UNCTAD) Africa Oil, Gas and Minerals Trade & Finance Conference and Exhibition held in STP in 2010 have reinforced the country's reputation as a prospective producer-state. Similarly, sporadic official announcements, assessments carried out by the IMF (2014) or the African Development Bank (African Development Fund n.d.), as well as media reports about the existence of considerable quantities of oil in STP's waters, plans to deploy advanced exploration technology in the JDZ, and recent exploration contracts for blocks in the EEZ have not ceased to sow optimism.[7] Yet many Santomeans have come to speak with disillusionment about their country's prospects. Hope for oil, according to Lourdes, a member of the Santomean parliament and a college teacher in her early sixties, has become a doubtful hope (*uma esperança duvidosa*).[8]

Lourdes' doubtful hope jars with oil's quasi-millenarian promise pronounced only a few years earlier. 'May God bless us all!' were Santomean President Fradique de Menezes' closing words at the ceremony marking the opening of corporate bids during STP's first licensing round for drilling rights in 2003. In total, signature bonuses amounting to around US$500 million had been offered, among them Chevron's highest bid of US$123 million for Block 1 of the JDZ, signalling imminent wealth. Divine providence and proper state management, together, were to usher in a new reality for the country.

A sense of millenarianism had already characterized the first business meetings with ERHC (Environment Remediation Holding Company), the little-known and largely inexperienced US oil exploration company that is said to have ushered in STP's petroleum era (*era do petróleo*) in the late 1990s. Afonso Varela, a Santomean lawyer who had attended some of these meetings, then in his capacity as an official of the Santomean Ministry of Finance, recounted the gung-ho attitude of the company representatives when we spoke in 2012. Oil, Varela recalled, was portrayed as something straightforward, something that would happen – if not overnight, then soon. The ERHC employees exuded confidence despite their lack of experience in the oil business and the riskiness of their investment. At this point in our conversation, Varela wavered, apparently questioning the certainty of his recollections: 'It is difficult to say whether it was the way [ERHC] presented it or whether it was [our] limited knowledge … that led us not to appreciate the problem fully'.

Thousands of barrels, valued at US$100 each, generating a million dollars or more a day – those were the simple but persuasive computations produced by simulations developed at World Bank workshops in the early 2000s. Cocoa's annual yield of US$3 to 4 million, as Varela remembered, quickly paled in comparison with those staggering numbers. Not all of these computations were made public, but they did not fail to affect. For while other African nations had experienced a post-independence boom based on generous resource endowments (Cooper 2002), STP had entered an economic crisis already by the time its colonizers left in 1975 (Hodges & Newitt 1988; Seibert 2006). It was aggravated by a drastic fall in cocoa prices,[9] a lack of agronomists, engineers, and planners to provide adequate management of the ailing plantations, and a seeming failure to synchronize the enthusiasm and aspirations of urban political cadres with those of marginalized plantation workers. Structural adjustment, introduced in the late 1980s and 1990s, did not halt the progressive deterioration of STP's agricultural base.

Petroleum, then, promised a break with the stringent impositions made by the international financial organizations, which sought to unmask anything that independence had achieved as pure illusion. Over a beer at Papa Figo's, a popular street café, friends would reminisce about the parties held three, four nights a week in the 1980s. A popular discothèque was the Bataclan, perched on a hill above the capital, near the military compound, and frequented not only by Santomeans but even by Angolans, who would fly in for an exciting night out. For Miguel, those times were the good times when he earned a monthly salary of STD6100, what today would be STD12 million (*c.*US$650). But it was money that turned out to be drawn from deficits and a mountain of debt. Soon, Miguel, like many others, would be labelled an 'overpaid urban worker' by the IMF (Cooper 2002: 116). STP was promised integration into global markets less through consumption than through debts of a different kind (cf. Mbembe & Roitman 1995: 336). The devaluation of the local currency, the dobra, in 1987 alongside the country's 'opening' (*abertura*) to global market forces increased foreign debt from US$23 million in 1980 to around US$144 million in 1990, amounting to more than 300 per cent of the country's GDP (Seibert 2006: 307).[10] The sense of abjection instilled by these events still echoed in Miguel's words (cf. Ferguson 1999: 236). 'What they live in Portugal today [following the Euro crisis]', he suggested when we talked in 2012, 'we already lived then. What they are crying about now, well, we've already experienced it'. The other men sitting with us around the table nodded in agreement.

Aires Bruzaca, a Santomean economist who had devoted his master's thesis to STP's oil, explained the situation to me as follows: 'When the idea of petroleum emerged, the country was surviving a situation of true … anxiety, looking for a solution to its problems – and petroleum was seen as a solution'. In 1990, a cheerful crowd had marched the returned political exile Miguel Trovoada from the airport into the city and made him President shortly afterwards, opening up the path to *mudança* (change) and multi-party democracy. However, many Santomeans were soon disappointed by *mudança*'s apparent inefficacy and by what Lourdes and her husband Carlos, two of Trovoada's erstwhile supporters, described as a game of competing interests and vote-buying, leaving the country without any serious perspective. For Santomeans, then, the announcement of oil, a few years later, furnished hopes for betterment with a concrete object – an object that, as Vincent Crapanzano (2004: 119) writes, seems to shorten hope's horizon, turning it into a more efficacious form of desire. 'Everybody thought they would receive their own barrel of *petróleo*' was a constant refrain I heard during my fieldwork, invoking images of personal and collective transformation enabled by access to fuel, increased mobility, and greater connectivity (Gardner 2012; Shever 2012).

In short, oil's promise has taken shape at the convergence of previous experiences of hope as well as disappointment mobilized in the present (Hemmings 2006: 564; Miyazaki 2004: 139). As a prospective national asset, oil has helped project a future ethical order where Santomeans might find redemption from a past constituted of derailed dreams and aspirations only half-attained. Not unlike earlier experiments with democracy and structural adjustment, oil has created expectations of prosperity and development, but this time with a possible degree of national sovereignty regained.

Significantly, however, the doubtful hope invoked by Lourdes has shifted oil's affect from a sudden surge that grabs you unexpectedly to something open to contestation and in need of careful handling. This hope is doubtful, first, about the right conditions that might allow oil to fulfil its promise and, second, about whether there will be any oil at all. Being in doubt is 'being of two minds' (Pelkmans 2013: 4). It constitutes not simply an act of questioning but, simultaneously, an attempt to remove or strengthen its own indeterminate foundations. Rather than giving way to outright disbelief, doubtful hope suggests actions by which uncertainty around petroleum's potential can be reduced, albeit with varying success.

The most prolific doubts about future oil in STP have revolved around the kind of wealth that oil will generate and whether it will be equitably distributed. Accusations of bribery and corruption in STP's oil sector abounded from the start, implicating even the popular former President de Menezes, who had repeatedly declared his unequivocal support for the implementation of transparency measures (Seibert 2006: 371, 377; Soares de Oliveira 2007: 233-42). Those measures comprised a spate of governance programmes and public campaigns that accompanied the three licensing rounds in 2003, 2004, and 2010, and aimed to increase accountability while sensitizing the population to oil's notoriously detrimental socioeconomic outcomes. Institutions such as the National Petroleum Agency, which manages and regulates STP's oil on behalf of the state, and a state-of-the-art oil revenue management law[11] embody proposals for an appropriate set of relations within which future oil needs to be embedded in order to bring prosperity for the greatest number. Alongside them, the campaigns for transparency by international

NGOs and global governance organizations play into widespread vigilance regarding the assumed vested interests of political leaders. Taken together, this work has contributed to the crystallization of a politics of anticipation, mobilizing particularly Santomean civil society groups, which seeks to engage and act upon an uncertain future with oil (Weszkalnys 2014).

Yet oil's contribution to the country's development has remained somewhat elusive. Seismic research, prospecting, and exploration activities have occurred offshore, hundreds of kilometres off STP's coastline. Little if any of the technical equipment and personnel has passed through the islands. To date, approximately US$60.5 million derived from signature bonuses for offshore acreage has been deposited in STP's national petroleum account, used largely, as stipulated by STP's oil revenue management law, for annual budget support (PWC 2015: 14).[12] In addition, the country has received an unspecified amount in the form of corporate support for social projects, including education facilities, scholarships, and sanitation infrastructures. Beyond these, there are few visible signs of change. No frantic stampede, no flights delivering oil workers from Houston or Port Harcourt, no multinationals constructing local headquarters, and no expat compounds being stomped out of the ground, as is happening, for example, in neighbouring Equatorial Guinea (Appel 2012). There are only a few new office blocks occupied by international insurance companies and banks that may, or may not, have been attracted by the prospect of oil.

In STP, as elsewhere, resource extraction has been characterized by a sense of opacity, provoking a steady stream of rumour and speculation (Gardner 2012; High 2013; Kirsch 2006; Limbert 2015; Walsh 2012; West 2006). If anything, this situation seems amplified by recent government statements in favour of a 'no oil' scenario. They have sought to downplay oil's potential, recommending alternative development strategies instead, including agriculture, industrial services, and tourism. In October 2011, the failure to invite members of the press to attend the signing ceremony for a contract with Nigerian firm Oranto Petroleum, which plans to explore block 3 of the EEZ, was widely interpreted as a failure to open the event to public scrutiny. Official explanations blamed oversight on the part of the technical staff in charge, and pointed to incumbent Prime Minister Trovoada's efforts to keep oil-related expectations in check. From another, more popular, perspective, however, the incident all but corroborated the government's effort to keep Santomeans in the dark about their oil. It was an instance of information, and hence power and wealth, not shared. A few months later, new rumours began to circulate regarding four large Taiwanese vessels that had anchored just off the Santomean coast. Although the vessels remained in plain sight for several months, the purpose of their presence was not revealed. When pressed on the issue, Trovoada cited potential business interests and plans to set up a floating port and transshipment hub by the vessels' owner, Bluesky-TMT.[13] Yet even the government officials who should have been managing the process seemed ignorant of the deal. A preferred explanation, therefore, offered by friends who recounted the incident later that year, invoked secret petroleum business with the Prime Minister pulling the strings. Indeed, the controversy around the Taiwanese vessels ultimately precipitated the censure motion brought successfully against Trovoada in November 2012.[14]

Confusion, doubt, and suspicion tell of the powerful politics of oil's affect. Just like high-spirited hope, they circulate not as autonomous forces but in ways shaped by institutional practices, official discourse, NGO campaigns, and incessant polemics. In one view, they provide prolific commentary on capitalist expansion and people's exclusion from its benefits (Gardner 2012; High 2013). Importantly, however, rather than indicating affect's escalation, they also reframe its object: oil. Doubt, from this perspective, implies an ethics of constraint (Crapanzano 2004: 100), which puts untrammelled hope at arm's length. Here, oil's opacity is not simply a metaphor. Rather, it emerges from a specific resource materiality, which doubt engages and seeks to negotiate.

A last example will suffice. During a return visit in November 2013, I encountered a new explanation for oil's continuing absence despite contracts, exploration zones, and test drills. It suggested that multinational companies were now holding Santomean oil 'in reserve' as they speculated on a rise in oil prices or a lowering of production costs through technological advances. This explanation was strengthened by French oil company Total's withdrawal from the JDZ a few months earlier – the last oil major that had held interests in STP's shared maritime territory with Nigeria. 'Do you believe we don't have any commercially viable petroleum?', my friend Silvino queried when we talked during this visit. Ever since we first met in 2007, Silvino had been an outspoken sceptic regarding the intentions of international oil companies operating in STP. Now he claimed to draw on information from a circle of unnamed, but well-informed, friends with special insight into the matter. He continued: 'Ghana to the north [of STP] has [oil], Angola to the south has it, and Nigeria and Gabon to the northeast'. For Silvino, proffering a theory featuring 'occult' global giants responsible for oil's continuing absence, the answer was obvious: 'We will only extract [oil] when they let us'.

One might argue that there is, indeed, very little that ordinary Santomeans can do to speed up the protracted process of oil exploration, even as its extended periods of apparent inactivity constitute a temporality difficult to reconcile with the country's everyday economic necessities. Whereas mineral rushes propelled by a multitude of individuals can open up new terrains for extraction, the enthusiasm and hope Santomeans have experienced in the face of oil will, in themselves, do little to drive petroleum activities. The capital-intensive nature of oil exploration and the requisite for specialized technology and know-how have resulted in dependency on foreign investments and international partners. Unsurprisingly, then, doubts about whether oil will come at all speculate, in the first instance, about STP's relations with its corporate collaborators and, more generally, the country's marginal position *vis-à-vis* global markets. From this perspective, the invocation of reserves by Silvino and others is also one of fugitive wealth, so familiar in the archipelago's history of resource exploitation. It seems remarkably cognizant of the inherent tensions in contemporary oil markets, where both the accounted-for presence and calculated absence of reserves can create value for corporations (cf. Limbert 2015). As some Santomeans would argue, the fact that the country *has* oil is already established. However, a fair distribution of the benefits has been hindered by a market logic that has enabled the concealment of oil's presence from the country's citizens. Yet few would spurn oil's promise of resource-related development – of jobs, infrastructures, and a transparent state – even if this promise has come to appear increasingly fragile.

Waiting or working for wealth?

In 2012 and 2013, I carried out a series of interviews with Santomean friends and acquaintances, in which I asked them to describe the moment when oil was first announced in STP. This was Lourdes' response:

> When they started talking about the possible existence of petroleum in São Tomé, everyone became euphoric, with many hopes that we would have a much better life. People began to prepare themselves. For example, they took English lessons, … because when petroleum comes that is the language you will need … We stayed with that hope, that expectation, for two years or so. The entire population stopped wanting to do anything because petroleum was coming. Everyone would have their own barrel, [people said] … Then, the expectation [of oil] was undermined … because petroleum didn't come. Meanwhile, one would hear that there were already people working in the petroleum offices, … and people saw [that there was] a group that was already receiving the money, … but [they themselves] weren't seeing any of it. So what happened? Agriculture was nearly abandoned, people gave up agriculture, there was a rural exodus, they came to the city and stayed here with the hope for oil.

To hope, *esperar*, is also to wait in Portuguese. Lourdes' account seems to play with this ambiguity. It takes me back to the question of how resource affect is being problematized. Lourdes maps out the trajectory I described in the previous pages: quasi-millenarian euphoria rapidly followed by doubts about the inequities oil extraction might magnify. There is, however, a third trope that figures centrally, namely that of waiting for oil. In a sense, Lourdes' words reflect the specific cruel optimism (Berlant 2006) of a nascent oil economy where people find themselves caught in between lives as exploration stagnates, sometimes for indefinite periods. But they also contain a popular moral critique of hopeful responses to oil inflected by class and categories of belonging in Santomean society, as well as long-standing discourses on work ethics and indolence. The worry here is not a possible outbreak of resource-induced frenzy and violence but a different kind of affective excess. In this last section, I want to begin to unpeel the complex genealogy of these interrelated notions: of hoping, waiting, and not working. I show how Lourdes' concerns articulate with a broader problematization of hopeful affect and non-productive labour in extractive economies.

Consider this warning in a recent United Nations Development Programme publication, entitled *Getting it right*, which sets out guidelines for the prudent management of hydrocarbon economies: 'It is not unusual for unreasonable expectations to be nurtured, with resources in the ground perceived as opening the door for all aspects of a "better life", almost as a large-scale collective lottery game' (UNDP 2011: 57). The UNDP publication is typical of a thriving body of work which highlights 'unreasonable expectations', specifically because they are feared to translate not into vibrant, energetic activity associated perhaps with a mining rush, but into its opposite. In particular, they are seen to 'undermine work ethics and distort perceptions of merit' (UNDP 2011: 42) as resource revenues seem to offer a guaranteed, effortless income and the ready satisfaction of demands placed by citizens on their governments. Hope for oil, in other words, is seen to find its corporeal expression in an explicit type of non-work: waiting. Waiting embodies an orientation of the self in time that, as Crapanzano has noted, is often stigmatized as the paralysis that comes with hope, when people wait 'passively for hope's object to occur' (2004: 114; see also 1986).[15] This constitutes a paradox from the perspective of affect theory, largely focused on the transitivity, unassimilability, and 'more than' of affective force. Hope, by contrast, is problematized as a temporal affect that repels the generativity and productiveness otherwise associated with the appropriation of resources as the foundation for economic wealth. Rather than implicating themselves productively as part of a future with oil, the citizens of emergent oil-producer nations are imagined to prefer to wait instead.

In STP, local assertions that Santomeans are now waiting for oil resonate with a common, pained self-judgement that 'Santomeans do not work'. Both, in turn, are reminiscent of a colonial discourse on native indolence. The latter reinforced a notion of an indisposition to appropriate the islands' rich agricultural potential as a characteristic of São Tomé's creole population. It was amplified, first, when the abolition of slavery in 1875 threatened to result in chronic labour shortages, which seemed to jeopardize the Portuguese colony's future, and by later periodic attempts, through legal obligations and taxation, to impose coercive forms of labour exploitation (Hodges & Newitt 1988; see also Whitehead 2000). The postcolonial period of structural adjustment created its own discourse of African idleness, seen to be embodied in public-sector sinecures and rent-seeking practices (Cooper 2002: 116). It continues to echo in Santomeans' common discontent with the state's overblown bureaucratic apparatus today. While the earlier, colonial stereotypes posited workshyness as the result of ecological, racial, or cultural conditions, twentieth-century critics largely blamed bad governance and a neo-patrimonial mentality. More recent accounts of hydrocarbon economies, by contrast, would consider Santomeans' alleged waiting for oil, and that of their counterparts in other African nations, a psychological effect triggered partly by the presence of valuable resources and by economic mismanagement. Though no longer applying explicit labels of laziness or indolence, the implications appear eerily similar.

What is remarkable about Santomean talk of oil-induced indolence and recent policy discourses and reports, such as *Getting it right*, is the shift away from questions of state governance and the conduct of political leaders to the ethics and expectations of the middle classes and poorer populations of resource-rich countries (e.g. Batega, Kiiza & Ssewanyana n.d.; Gilberthorpe *et al.* 2014; Kakonge 2011; Patey 2014; Samuel, Ernest & Ernest 2012). A substantial literature on extractive economies over the last two decades focused on a resource curse, resulting in part from the improper behaviour of national elites who devour resource wealth with impunity (Auty 1993; Karl 1997; see also Apter 2005; Coronil 1997). By contrast, the UNDP report now worries, more universally, about the supposed, dangerous transformation of 'the economically active population into rentiers' (2011: 46).

In an important sense, this generalizes long-standing unease about unearned and non-productive wealth derived from natural resources, which runs back to the eighteenth-century political economists and has seemingly continued into the present. Adam Smith, for example, anticipated these sentiments in *The wealth of nations* when deriding the Spaniards' unwise desire for South American gold. Mining projects, Smith reflected, 'are the projects ... to which of all others a prudent law-giver, who desired to increase the capital of his nation, would least choose to give any extraordinary encouragement' (1776: 215). Not only was the exploitation of such non-renewable resources considered to have the somewhat unique capacity to swallow both capital and profits; but rents, as David Ricardo noted, appeared to lack any underlying creative base. These political economists had come to understand value creation as the exclusive result of a productive process of labour acting on or enhancing supposedly inactive and subordinate natural matter (land, resources, raw materials). Rents, by contrast, rather than derived from human ingenuity or labour, were simply derived from nature itself.[16] Permutations of this theme are offered by Smith's and Ricardo's Portuguese contemporaries, condemning their royals' inability to turn the resources extracted from Brazil into more long-lasting wealth. Indulgence in pious luxury, including the construction of convents and lavish gifts to the See of Rome, and conspicuous consumption meant that '[t]he gold of Brazil merely passed through Portugal and cast

Journal of the Royal Anthropological Institute (N.S.), 127-146
© Royal Anthropological Institute 2016

anchor in England' (Oliveira Martins 1908, quoted in Hammond 1966: 8) and, one might add, the Vatican.

The imaginative connection I have sketched here between hoping, waiting, and not working is thrown into stark relief by notions of expectation management in contemporary hydrocarbon economies. So-called local content projects offer a particularly apposite example. The notion of local content epitomizes a development ethos focused on the creation of 'trickle-down' effects through broad-based participation in the extractive industries, for example, by employing local staff, subcontracting local businesses, and using their infrastructures for other, non-industry purposes (Tordo, Warner, Manzano & Anouti 2013). It aims to balance the problematic enclave character of mineral extraction (Ferguson 2005) and provides a state-led but explicitly market-based solution for a more equitable division of resource wealth between host states and industry, while bypassing reliance on government investments and what are now widely deemed the pitfalls of nationalization (Ferguson 2015: 172; Ovadia 2014; Soares de Oliveira 2007). Alongside macro-economic policies, institution building, transparency mechanisms, and public education, local content has become a favourite of contemporary expectations management. It suggests a productive involvement of the citizens of resource-rich countries in the creation of resource wealth. Put differently, rather than wait for windfalls from that 'large-scale collective lottery game' lamented in the *Getting it right* report, now people might be made to work for their share.

It is easy to identify some of local content's shortcomings without denigrating its potential to 'revers[e] decades of underdevelopment under global neoliberalism' (Ovadia 2014: 145). First, there is an obvious mismatch between the temporalities of politics and markets that are at stake. At a seminar on local content held in STP in 2007, international consultants revealed that, at present, the country lacked the human resources, infrastructures, and capital necessary to ensure meaningful participation (see Klueh *et al.* 2007; MundiServiços n.d.). Their painting of a rosy future where state policies, specialized training, and improved skills among the local business community could alleviate the problem seemed questionable to industry experts who attended the seminar. What is more, characteristic price volatility makes oil and other minerals a precarious economic foundation. Coupled with the marked slow-down of exploration activities in STP in recent years, any conceivable preparations for oil have taken on a peculiar sense of futility. Second, given STP's insularity and the extensive availability of industry-related services and resources elsewhere in the region, it seems reasonable to doubt the efficacy of local content policies. The material betterment that might be achieved by future Santomean oil production is likely to happen elsewhere. Third, though solid empirical measures of local content are still absent, there is some evidence that its policies are biased towards local elites (Ovadia 2014) or, in any case, only poorly address the sort of ecological and cosmological fractures generated by extractive industries discussed earlier in this paper.

Most striking, however, is local content's productionist premise, which appears increasingly ill suited to the economic landscape now presented by African states (Ferguson 2015). The overwhelming emphasis on job creation and productivity offers only a partial response to the underlying issue of wealth redistribution. It could do better at engaging the articulations between resource extraction and the different distributive mechanisms in resource-producer states, highlighted by James Ferguson, including state subsidies, patronage, kinship and other social networks, and future-generations funds of the type envisaged by STP's oil revenue management law. Local content advocates also rarely ask what kind of labour is carried out by those who are supposedly waiting, which might not correspond to conventional conceptions of productivity, or how resource windfalls compare with other non-productive forms of wealth creation, including the speculative labour of finance. Instead, local content policies cast the potential reliance on resource revenues as both politically problematic and morally reprehensible. They anticipate economic stagnation and societal breakdown, conflict, and violence, which might arise when hopes of instantaneous wealth remain unfulfilled. Local content thus targets citizens' expectations. Resource affect is both de-mobilized and re-mobilized with the aim of realizing new kinds of 'intensive socialities' (Anderson 2010: 170). Nothing less than an ethical reorientation is encouraged, substituting the wrong, futile, or excessive kind of hopes for the right ones.

Conclusion

In this paper, I have argued that affect has come to play an increasingly prominent, if somewhat nameless, role in global debates about natural resource extraction. We are witnessing a proliferation of concern situated at the intersection of corporate, government, and third-sector practice, which focuses no longer simply on macro-economic issues and elite politics but on the purported hopes, desires, and aspirations of citizens in producer states. As I discussed, affective responses to natural resources are not new. If mineral rushes show us how dispersed affective energies have helped open up extractive territories, this is now matched by a growing number of examples where the remonstrations and anger of those affected by extraction have successfully impeded resource exploitation in other locales. And while the cravings for power and personal wealth on the part of state officials no doubt continue to be among the key factors shaping the political and legal contours of extractive regimes, the Santomean case demonstrates that such processes are now subject to intense scrutiny. The concept of resource affect I have proposed allows anthropologists to theorize these multifaceted articulations of affect within contemporary resource economies, both as a highly problematized object and as an intrinsic element of capitalist dynamics.

Anthropologists' response to the emotional upheaval caused by resource booms has been to dispel exaggerated notions of human pathology, madness, and uncontrolled greed. They have pointed, for example, to the very genuine sense of belonging expressed by boomtown residents and their efforts to forge stable family lives (Rolston 2014; see also Ferguson 1999; Walsh 2012). I proposed a different tack: resource affect is less an externality of resource extraction to be explained away than constitutive of its social-material fabric. Formed from somatic intensities, preconscious yet 'of the crowd', resource affect consists of relations through which human beings and material substances are made and remade. Nor is resource affect simply a symptom of economic uncertainty, social inequity, geological and climatic conditions, or harmful government policies that have been named as causes for rushes in certain locales (Cleary 1990; MacMillan 1995). The sudden and impulsive temporality of the rush that grabs you, sometimes unawares, and its apparent opposite of doubting and waiting require us to look beyond the supposedly predictable economic oscillations of resource exploitation.

In STP, responses to oil have been both more varied and more discerning than simplistic notions of hope and expectation, characteristic of current policy discourse on natural resources, might imply. A multiplicity of affective registers have translated into a doubtful hope, comprised of ambiguous visions of material betterment, personal and collective transformation, as well as anticipations of failure, friction, and discontentment. People speculate about how their own lives might be transformed by oil, whether through higher salaries and employment opportunities, or electricity round the clock, paved roads, functioning sewerage systems, and new hospitals. Such concrete, material aspirations highlight what is now found wanting, and invoke an idea of an autonomous, sustainable state. In imagining what might be, my fieldwork companions drew on a variety of sources: from media reports to word-of-mouth, personal observations of corporate and government actions, stories collected of their neighbours along the Gulf of Guinea coast, and the campaigns of international NGOs. They were keenly aware that the fulfilment of their dreams could not be determined simply by the arrival of commercial oil production. As a result, oil is seen less as a panacea than as an object to be handled with care.

Santomeans' affective responses to resource extraction tell us something about the contradictory effects of contemporary capitalism and the ways people are caught up in them. They also speak to increasingly global concerns about how resource wealth may be shared equitably and how its uninterrupted production might be secured in the first place. In April 2014, I was invited to contribute to a workshop organized by the government of an emergent African oil-producer state and one of its corporate partners, an independent US oil company. A few years earlier, the same company had made a very successful oil discovery in a different country, ill prepared to deal with the consequences, and was now keen to apply the lessons learned. I and a number of other experts were to help delineate what our hosts called expectation management with examples of 'best practice'. The Although the term affect *per se* did not explicitly figure in our discussion, it alluded to a wide range of sentiments and grievances exhibited by a local population already aggravated by decades of regional and ethnic friction, in the face of future hydrocarbon extraction. Will such workshops alleviate the violence often wielded by resource extraction or merely extend the scope of governance to encompass the minutiae of citizens' affective lives? What seems clear is that, similar to the plans and projections routinely used by corporations in managing their labour force (Richard & Rudnyckyj 2009), emergent practices of expectation management in hydrocarbon economies are not simply representations of resource futures but constitutive of them. They manifest a new kind of extractive politics that, as I have argued, revolves around not simply the democratic and technical aspects of resource exploitation but increasingly their associated affective inconsistencies.

One aim of this paper has been to provide a critical angle on the question of why and how resources matter in our contemporary world – beyond their status as abstract commodities and repositories of wealth in global markets. Affect directs attention towards the articulation of uncertain futures, and the ways they intensify, fade, and circulate as resource prospects are predicted, pursued, and abandoned. Anthropologists are well positioned to chart the ways natural resources are apprehended for their generative *and* their destructive potentials – ecologically, cosmologically, economically, socially – and their capacity both to underpin and to challenge dominant orders. I have highlighted the shift towards the making of resource affect as problem field 'imbricated with multiple modes of power' (Anderson 2010: 183), including the explicit scrutinizing of its contours and content. This is not to detract from the ethnographic fact that resource affect comprises a range of historically and materially constituted capacities, forces, and dispositions, whose qualities vary when different substances are at stake. Rather than closing down the inquiry with this analytical gesture, it is an attempt to alert us to the challenging and ever-changing nature of our ethnographic inquiries.

NOTES

This paper is based on fieldwork sponsored by the British Academy and the John Fell Research Fund, and would have been inconceivable without the generous support of numerous individuals and institutions in STP. Earlier versions were presented at the Centre for Research on Socio-Cultural Change (CRESC), University of Manchester, in 2013, and in the Anthropology department at the London School of Economics and Political Science in 2014, and were further developed during a visiting research fellowship at the Zentrum Moderner Orient, Berlin, in 2015. Many of my LSE colleagues provided excellent comments along the way, especially Catherine Allerton, Rita Astuti, Laura Bear, Stephan Feuchtwang, Katy Gardner, Deborah James, Nick Long, Michael Scott, Charles Stafford, and Hans Steinmüller. I also thank Eeva Berglund, Tanya Richardson, and Gerhard Seibert for their invaluable critique. I am especially grateful for the patience and advice provided by Jessica Barnes and Andrew Mathews, as well as *JRAI*'s Editor and reviewers.

[1] My analytical concerns differ from Marx's notion of commodities' fetish character, which, though it includes some affective qualities, refers largely to the exchange value of things.

[2] I borrow from the title of a panel I convened with David M. Hughes at the meeting of the American Anthropological Association in 2013.

[3] World Bank, World Development Indicators, GDP 2014 (*http://data.worldbank.org/indicator/NY.GDP. MKTP.CD/countries?order=wbapi_data_value_2014%20wbapi_data_value%20wbapi_data_value-last&sort =asc&display=default*, accessed 11 January 2016).

[4] Many thanks to Nick Long for suggesting this term.

[5] I carried out twelve months of fieldwork between 2007 and 2008, with follow-up visits in 2009, 2012, and 2013.

[6] This was reinforced by public campaigns focused on the revitalization of the agricultural sector following the world food price crisis of 2007/8.

[7] 'Bloco 1 tem reserva de 100 milhões de barris de petróleo para ser explorado em 15 anos', Telá Nón, 2 December 2013 (*http://www.telanon.info/economia/2013/12/02/15087/bloco-1-tem-reserva-de-100-milhoes-de-barris-de-petroleo-para-ser-explorado-em-15-anos/*); 'Exploração conjunta de petróleo com a Nigéria dentre de 18 meses', RFI Português, 4 April 2014 (*http://pt.rfi.fr/africa/20140331-zona-de-exploracao-conjunta-de-petroleo-com-nigeria-nunca-esteve-tao-bem*); 'Equator Exploration conducts 3D seismic studies in São Tomé and Príncipe', Macauhub, 12 May 2015 (*http://www.macauhub.com.mo/en/2015/05/12/equator-exploration-conducts-3d-seismic-studies-in-sao-tome-and-principe/*); 'Galp, Kosmos awarded Sao Tome block', Rigzone, 27 October 2015 (*http://www.rigzone.com/news/oil_gas/a/141288/Galp_Kosmos_Awarded_Sao_Tome_Block*).

[8] Some of the personal names used in this paper have been changed.

[9] A sharp price slump occurred in 1979. Between 1973 and 1984, Santomean cocoa production fell from 11,587 tons to 3,378 tons (Hodges & Newitt 1988: 133).

[10] In 2007, the country benefited from substantial debt relief under the Highly Indebted Poor Countries (HIPC) initiative (US$314 million). However, as bilateral debt constitutes one of the government's main sources of finance, debt levels have again risen rapidly (75.8 per cent of GDP in 2011) (AfDB, OECD, UNDP, UNECA 2012).

[11] Adopted in 2004, it includes provisions for a national oil fund, prohibitions on borrowing against the country's oil resources, regional allocations, an oversight commission, and a public information office (Bell & Faria 2007).

[12] The signing of more recent contracts regarding the EEZ in October 2015 has brought in an additional US$2 million. However, the amount flowing into the account is only part of what STP has officially received in signature bonuses for the JDZ and EEZ. For example, substantial costs for the running of the authority managing the JDZ are deducted before payments are made to STP. The total amount received by STP is closer to $US95 million (Gerhard Seibert, pers. comm.).

[13] 'Barcos da empresa Bluesky terão encontrado "abrigo" em São Tomé na fuga da justiça internacional', Téla Nón, 6 June 2013 (*http://www.telanon.info/sociedade/2013/06/06/13434/barcos-da-empresa-bluesky-terao-encontrado-%E2%80%9Cabrigo%E2%80%9D-em-sao-tome-na-fuga-da-justica-internacional/*).

[14] Trovoada has since been remade Prime Minister after his party, ADI, won an absolute majority in the legislative elections in October 2014.

[15] My aim is less to ask what a notion of 'waiting' might tell us about an actual work ethic (if it tells us anything at all) than to examine its politicized uses and effects. Other scholars have questioned waiting's

connotations of passivity, viewing it instead as a specific temporal experience common to the powerless and those marginalized by globalization, but also as potentially creative (e.g. Auyero 2012; Jeffrey & Young 2012).

[16] Ricardo was critical of Smith's colloquial conception of rents. He argued that rents should be understood more narrowly as 'that compensation, which is paid to the owner of land for the use of its original and indestructible powers' (2001 [1817]: 40). Revenues derived from the removal of above- or below-ground resources, such as timber or minerals, do not constitute rents in this sense.

REFERENCES

AfDB, OECD, UNDP, UNECA 2012. African Economic Outlook – São Tomé and Príncipe 2012 (available on-line: *http://www.afdb.org/fileadmin/uploads/afdb/Documents/Publications/Sao%20Tom%C3%A9%20and%20Principe%20Full%20PDF%20Country%20Note_01.pdf*, accessed 19 January 2016).

AFRICAN DEVELOPMENT FUND n.d. São Tomé and Príncipe: maximizing oil wealth for equitable growth and sustainable socio-economic development. Abidjan: African Development Bank (available on-line: *http://www.afdb.org/fileadmin/uploads/afdb/Documents/Project-and-Operations/Sao%20Tome%20and%20Principe%20-%20Maximizing%20oil%20wealth%20for%20equitable%20growth%20and%20sustainable%20socio-economic%20development.pdf*, accessed 19 January 2016).

ANDERSON, B. 2010. Modulating the excess of affect: morale in a state of 'total war'. In *The affect theory reader* (eds) M. Gregg & G.J. Seigworth, 161-85. Durham, N.C.: Duke University Press.

APPEL, H. 2012. Walls and white elephants: oil extraction, responsibility, and infrastructural violence in Equatorial Guinea. *Ethnography* **13**, 439-65.

APTER, A. 2005. *The Pan-African nation: oil and the spectacle of culture in Nigeria.* Chicago: University Press.

AUTY, R. 1993. *Sustaining development in mineral economies: the resource curse thesis.* London: Routledge.

AUYERO, J. 2012. *Patients of the state: the politics of waiting in Argentina.* Durham, N.C.: Duke University Press.

BATEGA, L., J. KIIZA & S. SSEWANYANA n.d. *Oil discovery in Uganda: managing expectations.* Kampala: Economic Policy Research Centre and Makerere University.

BEHRENDS, A. 2008. Fighting for oil when there is no oil yet: the Darfur-Chad border. *Focaal* **52**, 39-56.

BELL, J.C. & T.M. FARIA 2007. Critical issues for a revenue management law. In *Escaping the resource curse* (eds) M. Humphreys, J. Sachs & J. Stiglitz, 286-321. New York: Columbia University Press.

BERLANT, L. 2006. Cruel optimism. *Differences* **17**, 20-36.

BLACKMAN, L. 2012. *Immaterial bodies: affect, embodiment, mediation.* London: Sage.

BRYCESON, D.F., J.B. JONSSON & R. SHERRINGTON 2010. Miners' magic: artisanal mining, the albino fetish and murder in Tanzania. *Journal of Modern African Studies* **48**, 353-82.

CLARENCE-SMITH, W.G. 2000. *Cocoa and chocolate, 1765-1914.* London: Routledge.

CLEARY, D. 1990. *Anatomy of the Amazon gold rush.* Basingstoke: Macmillan.

COOPER, F. 2002. *Africa since 1940: the past of the present.* Cambridge: University Press.

CORONIL, F. 1997. *The magical state: nature, money, and modernity in Venezuela.* Chicago: University Press.

CRAPANZANO, V. 1986. *Waiting: the whites of South Africa.* New York: Vintage.

——— 2004. *Imaginative horizons: an essay in literary-philosophical anthropology.* Chicago: University Press.

FERGUSON, J. 1999. *Expectations of modernity: myths and meanings of urban life on the Zambian Copperbelt.* Berkeley: University of California Press.

——— 2005. Seeing like an oil company: space, security, and global capital in neoliberal Africa. *American Anthropologist* **107**, 377-82.

——— 2015. *Give a man a fish: reflections on the new politics of distribution.* Durham, N.C.: Duke University Press.

FERRY, E.E. & M.E. LIMBERT 2008. Introduction. In *Timely assets: the politics of resources and their temporalities* (eds) E.E. Ferry & M.E. Limbert, 3-24. Santa Fe, N.M.: School for Advanced Research Press.

FOUCAULT, M. 2000. Polemics, politics and problematizations. In *Ethics: subjectivity and truth (Essential works of Michel Foucault 1954-1984)* (ed. P. Rabinow), 111-20. London: Penguin.

GARDNER, K. 2012. *Discordant development: global capitalism and the struggle for connection in Bangladesh.* London: Pluto.

GILBERTHORPE, E., S.F. CLARKE & P. SILLITOE 2014. Money rain: the resource curse in two oil and gas economies. In *Sustainable development: an appraisal from the Gulf region* (ed.) P. Sillitoe, 153-77. Oxford: Berghahn.

HAMMOND, R.J. 1966. *Portugal and Africa 1815-1910: a study in uneconomic imperialism.* Stanford: University Press.

HEMMINGS, C. 2006. Invoking affect. *Cultural Studies* **19**, 548-67.

HIGH, M.M. 2013. Believing in spirits, doubting the cosmos: religious reflexivity in the Mongolian gold mines. In *Ethnographies of doubt: faith and uncertainty in contemporary societies* (ed.) M. Pelkmans, 59-84. London: I.B. Tauris.

HODGES, T. & M. NEWITT 1988. *São Tomé and Príncipe: from plantation colony to micro-state.* Boulder, Colo.: Westview.

IMF 2014. Democratic Republic of São Tomé and Príncipe: Poverty Reduction Strategy Paper. Washington, D.C.: International Monetary Fund.

JEFFREY, C. & S. YOUNG 2012. Waiting for change: youth, caste and politics in India. *Economy and Society* **41**, 638-61.

JORGENSEN, D. 2006. Hinterland history: the Ok Tedi mine and its cultural consequences in Telefolmin. *The Contemporary Pacific* **18**, 233-63.

KAKONGE, J.O. 2011. Challenges of managing expectations of newly emerging oil and gas producers of the South. *Journal of World Energy Law and Business* **4**, 124-35.

KARL, T.L. 1997. *The paradox of plenty: oil booms and petro states.* Berkeley: University of California Press.

KIRSCH, S. 2006. *Reverse anthropology: indigenous analysis of social and environmental relations in New Guinea.* Stanford: University Press.

KLUEH, U., G. PASTOR, A. SEGURA & W. ZARATE 2007. Inter-sectoral linkages and local content in extractive industries and beyond – the case of São Tomé and Príncipe (Working Paper WP/07/2013). Washington, D.C.: International Monetary Fund.

LIMBERT, M. 2015. Reserves, secrecy, and the science of oil prognostication in Southern Arabia. In *Subterranean estates: lifeworlds of oil and gas* (eds) H. Appel, A. Mason & M. Watts, 340-53. Ithaca, N.Y.: Cornell University Press.

MACMILLAN, G. 1995. *At the end of the rainbow? Gold, land, and people in the Brazilian Amazon.* London: Earthscan Publications.

MASSUMI, B. 2002. *Parables of the virtual: movement, affect, sensation.* Durham, N.C.: Duke University Press.

MAZZARELLA, W. 2009. Affect: what is it good for? In *Enchantments of modernity: empire, nation, globalization* (ed.) S. Dube, 291-309. London: Routledge.

MBEMBE, A. & J. ROITMAN 1995. Figures of the subject in times of crisis. *Public Culture* **7**, 323-52.

MIYAZAKI, H. 2004. *The method of hope: anthropology, philosophy, and Fijian knowledge.* Stanford: University Press.

MUNDISERVIÇOS n.d. Uma nova era de desenvolvimento e riqueza. Unpublished report. Lisbon.

NASH, J. 1993 [1979]. *We eat the mines and the mines eat us: dependency and exploitation in Bolivian tin mines.* New York: Columbia University Press.

NAVARO-YASHIN, Y. 2012. *The make-believe state: affective geography in a postwar polity.* Durham, N.C.: Duke University Press.

OLIVEIRA MARTINS, J.P. 1908. *História de Portugal,* vol. 2. Lisbon: Parceria A.M. Pereira.

OVADIA, J.S. 2014. Local content and natural resource governance: the cases of Angola and Nigeria. *The Extractive Industries and Society* **1**, 137-46.

PATEY, L. 2014. Kenya: an African oil upstart in transition (OIES paper: WPM 53). Oxford: Oxford Institute for Energy Studies.

PAULL, R. 1963. *Old Walhalla: portrait of a gold town.* Melbourne: University Press.

PELKMANS, M. 2013. Outline for an ethnography of doubt. In *Ethnographies of doubt: faith and uncertainty in contemporary societies* (ed.) M. Pelkmans, 1-42. London: I.B. Tauris.

PWC 2015. São Tomé and Príncipe. Second EITI report 2014. Lisbon.

RICARDO, D. 2001 [1817]. *On the principles of political economy and taxation.* Kitchener, Ont.: Batoche Books.

RICHARD, A. & D. RUDNYCKYJ 2009. Economies of affect. *Journal of the Royal Anthropological Institute* **15**, 57-77.

RICHARDSON, T. & G. WESZKALNYS 2014. Resource materialities. *Anthropological Quarterly* **87**, 5-30.

ROLSTON, J.S. 2014. Specters of syndromes and the everyday lives of Wyoming energy workers. In *Cultures of energy: power, practices, technologies* (eds) S. Strauss, S. Rupp & T. Love, 213-26. Walnut Creek, Calif.: Left Coast Press.

SAMUEL, Y.A., K. ERNEST & K. ERNEST 2012. Empirical assessment of expectations associated with the recent discovery of commercialisable oil in Ghana. *International Review of Management and Marketing* **2**, 177-91.

SEIBERT, G. 2006. *Comrades, clients and cousins: colonialism, socialism and democratization in São Tomé and Príncipe.* Leiden: Brill.

——— 2015. São Tomé and Príncipe economy. In *Africa south of the Sahara 2016* (ed.) Europa Publications, 1010-17. London: Routledge.

Shever, E. 2012. *Resources for reform: oil and neoliberalism in Argentina*. Stanford: University Press.

Smith, A. 1776. *An inquiry into the nature and causes of the wealth of nations*. London: W. Strahan & T. Cadell.

Soares de Oliveira, R. 2007. *Oil and politics in the Gulf of Guinea*. London: Hurst & Co.

Taussig, M. 1980. *The devil and commodity fetishism in South America*. Chapel Hill: University of North Carolina Press.

Tordo, S., M. Warner, O.E. Manzano & Y. Anouti 2013. *Local content policies in the oil and gas sector*. Washington, D.C.: The World Bank.

Tsing, A.L. 2005. *Friction: an ethnography of global connection*. Princeton: University Press.

UNDP 2011. *Getting it right: lessons from the South in managing hydrocarbon economies*. New York: United Nations Development Programme.

Walsh, A. 2003. 'Hot money' and daring consumption in a northern Malagasy sapphire-mining town. *American Ethnologist* **30**, 290-305.

——— 2012. After the rush: living with uncertainty in a Malagasy mining town. *Africa* **82**, 235-51.

West, P. 2005. Translation, value, and space: theorizing an ethnographic and engaged environmental anthropology. *American Anthropologist* **107**, 632-42.

——— 2006. Environmental conservation and mining: between experience and expectation in the Eastern Highlands of Papua New Guinea. *The Contemporary Pacific* **18**, 295-313.

Weszkalnys, G. 2011. Cursed resources, or articulations of economic theory in the Gulf of Guinea. *Economy and Society* **40**, 345-72.

——— 2014. Anticipating oil: the temporal politics of a disaster yet to come. *The Sociological Review* **62**, 211-35.

——— 2015. Geology, potentiality, speculation: on the indeterminacy of 'first oil'. *Cultural Anthropology* **30**, 611-39.

Whitehead, A. 2000. Continuities and discontinuities in political constructions of the working man in rural Sub-Saharan Africa: the 'lazy man' in African agriculture. *European Journal of Development Research* **12**, 23-52.

Yanagisako, S. 2012. Immaterial and industrial labor: on false binaries in Hardt and Negri's trilogy. *Focaal* **64**, 16-23.

Zaloom, C. 2009. How to read the future: the yield curve, affect, and financial prediction. *Public Culture* **21**, 245-68.

L'espoir et le doute : affect et ressources dans une future économie pétrolière

Résumé

L'affect joue un rôle de plus en plus présent, quoique rarement nommé, dans les débats mondiaux sur l'extraction des ressources naturelles. Cet article propose de théoriser l'affect lié aux ressources à la fois comme un élément intrinsèque de la dynamique capitaliste et comme un objet transformé en problème par la pratique des entreprises, des gouvernements et du secteur associatif. Sur la base de recherches ethnographiques à São Tomé et Príncipe, l'auteure explore les horizons affectifs auxquels la perspective d'une prospection pétrolière et gazière donne naissance : un espoir dubitatif, composé à la fois de visions d'améliorations matérielles, de transformations personnelles et collectives et de prédictions d'échec, de heurts et de mécontentement. Elle examine également la multitude de campagnes, activités et programmes lancés à São Tomé et Príncipe par des organisations non gouvernementales et des institutions de gouvernance mondiale autour de la question du pétrole, et des interrogations spécifiques que soulève le futur de la ressource. À la lumière de cet examen, elle avance que l'on voit émerger une nouvelle politique des ressources axée non pas simplement sur les aspects démocratiques et technique de leur exploitation, mais de plus en plus sur les dissonances et discordances affectives qui leur sont associées.

8

Liquid Oman: oil, water, and causality in Southern Arabia

MANDANA LIMBERT *City University of New York*

This paper explores how different natural resources figure in temporal imaginings. I ask: how do oil and water come to frame the relationships and chronologies of transformation and, more particularly, of causality? In order to understand visions of environmental futures, not only might we need to attend to the forms of planning, expectation, and prognosis that shape knowledge or senses of the future, but we may also consider how and why causality and significant events are associated with particular natural resources. Drawing on my previous work that explores the future orientation of oil-depletion talk in Oman as well as textual sources and ethnography, I argue that while water has been associated with pious rule and divine presence, oil has been considered to be much more transitory and the product of human interventions and policies, interventions and policies that emerge from a fraught political history. While water seems to motivate events and cause change, serving as an indication and vehicle of God's power, oil appears less a cause of national transformation, at least in its origins.

On 8 June 2010, the English-language newspaper *The Muscat Daily* reprinted an opinion piece by Kamal Sultan, one of a handful of outspoken critics of economic policy in the Sultanate of Oman.[1] The piece, entitled 'When our oil is exhausted', publicly articulated a common and continuing concern in Oman with the future depletion of its oil reserves and the country's attempt to grapple with economic diversification. Official statements from government spokesmen and oil industry engineers and executives, comments by many people outside the oil industry, and newspaper articles in the foreign press predict that oil will be depleted in Oman in twenty years. Though just one among many articles in both the foreign and national press about Oman's future hardship, this particular piece – as a reprint from at least ten years prior – also highlighted the continuing refrain and deferred quality of Oman's depletion projections. Since the late 1960s, when commercial amounts of oil began to be exported, the prediction of Omani oil industry officials, repeated popularly in the country and outside, has been that oil will be depleted within this same twenty-year horizon.

However deferred, this future orientation of Oman's oil depletion is also significantly different from the way oil appears in historical and national representations in many other petro-states, where images of gushers and rigs in state symbols and national

Journal of the Royal Anthropological Institute (N.S.), 147-162
© Royal Anthropological Institute 2016

historiography mark aspirational hopes for the future and narratives of anti-colonial struggles for self-determination and self-sufficiency in the past.[2] Instead, in Oman, oil's *discovery* is barely mentioned, either in everyday discussions or in accounts of the making of the modern nation-state. What is striking in Oman, therefore, is not only the way that oil appears so consistently and repeatedly in accounts of potential difficulties in the future, but also how oil's arrival and origins are quite absent in national and local representations of the country's modernity and prosperity.

I have argued elsewhere (Limbert 2008; 2010; 2015) that the constant and continuing debates and expectations about future oil depletion have produced a particular framing of Oman's present and its successful modernity as temporary and an interlude, an anomalous in-between time of prosperity. Unlike other developmentalist states, where the future is said to promise continued progress and greater prosperity, based on an 'expectation of permanence' (Ferguson 1999), in Oman, developmentalist discourse is distinctly tempered and its modernity often described as transitory. Such temporariness, I have also suggested, helps shape the *present* (rather than the past or the future) as Oman's golden age, guided and enabled not so much by oil as by the wise leadership of the current Sultan.

This paper draws on my previous work, but also explores the silences around oil's exploration and discovery. At the same time, this paper contrasts representations of oil's role in Oman's historical transformations with the ways that water appears in Omani historiography and historical talk. While oil appears as barely transformative in its discovery, and yet the primary cause of future hardship in its ultimate depletion, water is not only described as a harbinger of divine pleasure and anger, but is also a miraculous motivator and actor. How do different natural resources, therefore, figure in temporal imaginings? How do they frame the relationships and chronologies of transformation and, more particularly, of causality? In order to understand visions of environmental futures, not only might we need to attend to the forms of planning, expectation, and prognosis that shape knowledge or senses of the future, but we may also consider how and why causality and significant events are associated with particular natural resources. I argue that while water has been associated with pious rule and divine presence, oil has been considered to be much more transitory and the product of human interventions and policies, interventions and policies that emerge from a rather fraught political history. While water seems to motivate events and cause change, serving as an indication and vehicle of God's power, oil appears less a cause of national transformation, at least in its origins.

As James Laidlaw (2010) has argued in his analysis of the concept of agency in anthropology, 'the determination of what is and is not causally significant is not, as the rhetoric of ANT [Actor-Network Theory] sometimes suggests, a straightforwardly factual matter but a matter of interpretation' (2010: 146). As opposed to the practice-theory approach to agency, which 'smuggles rather specific values into a concept of the individual's efficaciousness [and] imagines this to consist of a creative force deriving from the interior of the human individual' (2010: 145), Laidlaw points out that with ANT an actor is any entity that plays a relatively independent causal role in bringing about change, whether this actor is human or nonhuman, animate or inanimate. While this ANT approach does not depend on consciousness or subjectivity, it also circumvents the question of how, why, or whether different people or political forces frame entities as causing change. Here, I illustrate Laidlaw's critical point further by examining the ways that the respective causalities associated with oil and water have appeared differently in Omani national historiography and everyday discussions of national formation. In particular, by exploring the events that these natural resources are said to propel (or not), I consider how responsibility and cause are allocated. While water has been associated firmly with divine and miraculous power, oil has been more squarely a human-mediated resource (even if ultimately oil cannot be purely human either), where the discovery, presence, and extractive infrastructure become less awe inspiring, and indeed embedded in a contentious past.

With regard to events, we may remember that within anthropology, as Bruce Kapferer has described, they have served exceptionally well as exemplary moments in the analysis of culture, politics, and social relations, or, as with the Manchester school and the work of Max Gluckman, as sites of crisis and as 'playing through of conflict' (Kapferer 2010: 6). Events, here, often reveal the limitations of normative patterns and expectations. Beyond their ethnographic use in social analysis, events are also, of course, often considered to be key elements of narratives, whether historical or fictional, serving as significant moments of transformation or revelation. This paper draws on this second approach to events. That is, rather than taking a moment from fieldwork experience that the ethnographer then analyses to elucidate a general pattern or to explore social tensions, it examines how particular instances in time (or sequences) become, or do not become, understood or framed as events, either large or small. 'Eventilization', as Lauren Berlant calls it, becomes significant here as it encourages attention not only to the ways that events force a shift in situations (as Alain Badiou [2001] points out), but also to how a 'process will eventually appear monumentally as form' (Berlant 2008: 5) and singular. Here, however, my focus on the process of becoming of an event, or not, is not to analyse a traumatic present or a crisis lived with ordinariness, but to consider how change and transformation – the separating of times into before and after, as Caroline Humphrey (2008) also notes – are themselves understood, and, further, how causality and responsibility for such events are allocated, whether in the past, present, or future.[3]

In so doing, we may recognize the different interpretative frames and sensibilities associated with arguably similar substances: 'natural' liquids that help produce wealth and prosperity, however unequally distributed that wealth may be. How are processes pertaining to oil and water, therefore, understood or framed as events, or not, and thus what might these resources be seen to cause, or have caused, in the past or future? Indeed, what are the potential effects of the ways that oil and water are framed as causing change in Omani historiography and talk about the future?

The non-cause and the non-event of oil: discovery

Compared to other states of the Arabian Peninsula, oil was discovered rather late in Oman. Though the first concessions were given to the D'Arcy Corporation in the 1920s, commercial amounts of oil were not discovered until the mid-1960s.[4] Indeed, it was not until 1967 that Oman exported its first commercial cargo.

It is not, however, the relative late discovery and exportation of oil that seems to permeate its narrativization in Omani historiography. Rather, what is striking is the discovery's diluted presence in national narratives. Most likely, its lack of presence reflects the highly contentious religio-political environment in which oil was discovered, drilled, and exported. As memories of those environments are sustained in some texts and everyday references by elders in interior Oman, those who may oppose state presence in the region can also call upon these histories, rooted as they are in a history of the existence of a separate theocratic state (the Imamate) through the 1950s, for future political mobilization. To understand the contours of the tensions surrounding territory, oil, and power, a brief description of what has come to be known as the Jebel Akhdar War is necessary. Though the war ended in the late 1950s (with guerrilla fighting lasting into the 1960s), it has continued to reverberate in everyday political and religious life. The war's connection to oil exploration is undeniable, as it was the arrival of exploration teams with military escort and the military control of towns *en route* to potential oil fields that served as the catalyst of the war.

Until the 1950s and 1960s, the territory where oil was ultimately discovered was controlled not by the al-Bu Saidi Sultan (whose centre of power was in the capital Muscat on the coast and who maintained a close – though often difficult relationship – with British officials), but by a theocratic figure, an Imam, who led a religio-political administration known as the Imamate in what is now the 'Interior Region' (*al-Dakhiliya*) of Oman. The Imamate, based on Ibadism (a third branch of Islam after Sunnism and Shi'ism and the branch of Islam dominant in this region), had been re-established in 1913 after a hiatus of about fifty years. In Ibadism, the Imamate exists in four potential states of religion: secrecy, sacrifice, defence, and manifestation.[5] Under particular conditions and with the availability of an appropriate candidate, a group of scholars can meet and select an Imam, which is what happened in 1913. As might be expected, much has been written and debated within Ibadi scholarship on the conditions of the selection, the qualities of the person, and the potential deposition of such a figure. Suffice it to say, an Ibadi revival was well established by 1913, influenced by other religious revival movements in North Africa, the Levant, the Arabian Peninsula, and the Indian Ocean as well as spurred by opposition to the British presence in Muscat. The British presence in Zanzibar, which had been under Omani rule until it became a British protectorate (though still an 'Arab Sultanate') in 1890, was also a factor in the revival.

In 1914, the most dogmatic and outspoken of the pro-revival figures, Nur al-Din al-Salimi, died, and by 1920 so had the Imam, Salim bin Rashid al-Kharusi. With the influence of other advisers, the new Imam, Mohammad bin Abdullah al-Khalili, signed an agreement with the Sultan of Muscat, known as the Treaty of Sīb, that in effect recognized the coexistence of the two governments: the Imamate in the interior and the Sultanate on the coast.

Though the thirty-year period that followed was one of relative peace between the Imamate and the Sultanate, there is little doubt that the Imams and their close advisers were well aware of and worried about the possibility that the Sultan would try to gain interior territory for unification. The question, of course, was when and how he would do so. For his part, Sultan Said bin Taimur was also aware that any movement into Imamate territory from his associates, British officials, or staff from the Iraq Petroleum Company (IPC) would not be welcome. Though it seems that the Sultan was intent on unification, he was also sensitive to the potential intensity of the opposition. He insisted that an IPC expedition needed military escort and he was reluctant, according to John Peterson, to allow any exploration beyond the coastal strip (2007: 57-62).

Tensions between the Sultan and the interior escalated with Imam al-Khalili's demise, the election of a third twentieth-century Imam, Ghalib bin Ali al-Hinai, in 1954, and the decision by the IPC to pursue oil exploration at Fahud, within Imamate territory. The ease with which the Sultan's forces took the town of Ibri, on their way to Fahud, in 1954 seemed to suggest a smooth offensive and indeed, despite occasional tensions, the following year also saw the surrender of the town of Nizwa, the presumed capital of the

Imamate, to the Sultanate. By December 1955, both Ghalib bin Ali and then his brother, Talib bin Ali (who was well known to be the real power behind the Imamate), had fled to Saudi Arabia, and, despite occasional battles, the Sultan's flag continued to fly over the forts of interior Oman through 1956. In the meantime, however, Imamate supporters in Saudi Arabia were gathering strength, preparing their own attack in May/June 1957. In July that year, the Sultan significantly lost control of the interior, but then regained it at the end of the month with the bombing of the towns and forts, including those of Tanuf and Bahla, the town where I have conducted research since the mid-1990s.

Personal memory of that day in Bahla at the end of July 1957, though changing, of course, was something readily recalled until recently. During fieldwork in the 1990s, friends regularly told me of the day the fort was bombed, as well as the previous day – or was it two? – when pamphlets warning and urging surrender 'fell from the sky'. People showed me where their parents hid weapons in their homes in preparation for battle, and what field they were in when the bombing started. As July bled into August, British ground troops met some resistance in Nizwa, but more in Bahla. A guerrilla war developed within the Jebel Akhdar mountains, but the major towns were captured and controlled. The war in the mountains continued for years, with support and sympathy from foreign states and other revolutionary groups as well as from individuals from the towns and villages of the interior who smuggled weapons from other Gulf territories and ventured into the mountains to fight.

The war eventually ended with the collapse of the Imamate administration and the integration of its territory into the Sultan's on the coast. At first, this new polity was known as the Sultanate of Muscat *and* Oman, joining the two quite separate political entities, but after 1970, when Sultan Qaboos bin Said al-Bu Saidi replaced his father as Sultan, the country came simply to be known as the Sultanate of Oman.

This war and the history of the Imamate continue to serve as reminders of alternative political arrangements to those currently in place as well as a past of tensions and animosities between different religio-political systems in Southern Arabia. The ethical and pious weight of this overthrown theocratic regime also serves as a counterpoint, articulated sometimes explicitly but often in subtle ways, to questions of contemporary legitimacy and practices of rule and corruption.

It is therefore no surprise to see few references to the war and the twentieth-century instances of theocratic rule in national histories. Indeed, the particular political situation of the lead-up to and events of the Jebel Akhdar war appear in only oblique references to 'logistical problems' in the official accounts of oil discovery. The official site of the Ministry of Oil and Gas is an excellent example:

> Until 1954, the exploration activities were limited to geological studies. This was due to several challenges, including those related to logistics, which made access to the most promising oil prospects difficult knowing the limitation in roads & transportation. The effort in search for oil continued and in 1955, DCSC [Dhofar City Services Company] drilled the first wildcat well in south Oman, named Douqah-1 and in 1956, PDO [Petroleum Development Oman] drilled its first exploration well in north Oman Fahud, named FHD-1. Both wells were dry. These failures combined with worsening logistical problems and a surplus oil supply in the world market led to withdrawal of some of the oil companies as operators (Ministry of Oil and Gas n.d.: 2–3).

Stories of discovery and export such as this are decidedly under-represented, or, rather, their presence in Omani historiography is decidedly understated. There are no heroes, no accounts of difficult terrain (besides 'logistical' ones), no sleepless and stressful nights, or exciting surprises. There is also no glorification of the shared liquid beneath 'our' land, nor are there struggles over the rightful national ownership of this liquid stolen by greedy multinationals or through neo-colonial tactics. There are no images of gushers in the telling of Oman's emergence as a modern nation-state in national museums, no images of oil rigs blanketing public spaces, and no state-sponsored documentaries that glorify the official opening of oil fields.

Official histories barely mention oil, and even school textbooks in Oman heralding the establishment of the modern nation only tangentially mention it. Nor have I even heard the discovery of oil as its opposite, a *curse*, as in the famous words of one of the founders of OPEC: that oil is the excrement of the earth. Indeed, when first researching Oman's oil history, it was surprisingly difficult to find information, including basic facts about concessions and dates, in *any* secondary Omani historical accounts. This was a project for archival work at the India Office Library. Therefore, unlike in most petro-states, where such histories are emphasized, the discovery of oil or its connection to the social body in Oman does not provide the rubric for the nation's narrativization. The national body is not brought together or defined through its shared access to oil or oil wealth, nor is the discovery of oil a defining event in its transition to modernity or a new formation to be celebrated. In other words, this is not Venezuela, Iran, the United Arab Emirates (UAE), or Saudi Arabia.

This does not mean that there is no celebratory emergence into a national modernity. On the contrary, Omani history has been structured precisely around a break when modern Oman begins, that is, with the transfer of power from father to son in a *coup d'état* in 1970, launching an era with its own proper name: *al-Nahda*, renaissance or renewal. All historical texts published after 1970 as well as most everyday conversations about the current conditions in Oman, official and unofficial, local and foreign, mention the great change after the July 1970 *coup d'état*. The relationship between the current Sultan and oil is usually described as one in which he brought 'wise leadership', using the country's wealth (i.e. oil) for the benefit of all. Thus, the *cause* of the national transition to modernity becomes not the discovery of oil, but the arrival of a new leader who harnessed an already-existing wealth.

In school textbooks and more recent official histories, oil also appears in very understated forms. A junior high-school social studies textbook, for example, which outlines the customs and traditions of Oman and includes oil, does so in the same kind of language described above. Among chapters on boat-building, pottery, goat-herding, weaving and looms, manuscripts, a famous Oryx project, and archaeological finds, we find a chapter on the discovery and exploration of oil. On the one hand, by incorporating oil into customs and traditions, the textbook may be understood as including it as part of the nation's cultural heritage. On the other hand, however, the language is similar to the Ministry of Oil's description provided above, with an account of oil that simply outlines the dates of concessions, the passing of one concession to another, the initial discovery of oil that was too heavy for export, the eventual discovery of usable, lighter, oil, the laying of the pipes, the first exportation, and, finally, a section providing the average number of barrels per day produced over the years. The language is direct, unembellished, and unglorified. Similarly, not surprisingly, neither is there mention of the wars that oil exploration caused or the Marxist rebellion in the southern region of Dhofar, which was also sometimes structured around claims about oil companies.

As in this textbook, in most officially sanctioned historical texts written in the 1990s and early 2000s, the wars of the emergence of the state – the Jebel Akhdar war, mentioned above, as well as the Marxist Dhofar war in Southern Oman and the Bureimi war between Saudi Arabia, the UAE, and Oman – barely appear. But such wars begin to appear more regularly in the late 2000s. For example, *Al-tarbīyat al-mūwāṭinīya fī fikr al-Sulṭān Qābūs* (National education in the thought of Sultan Qaboos) begins, like most contemporary texts, with an introductory section on general Omani history. Here, however, rather than a chronology of rulers, their dates, and the highlights of their rule (often silent about any political movement or past that may have contemporary resonance), there is a very brief account of the twentieth century as encompassing a series of wars, including the ones just mentioned, and a brief account of the discovery of oil. Oil appears in this Omani history simply as such: 'But after some positive occurrences for the Sultanate, there was the problem of Bureimi between the Sultanate and the Saudi Kingdom, [then] oil (*al-naft*) was discovered (*iktishāf*) and began to be exported (*bada' taṣdīr*) in 1967' (Ministry of Education 2007: 23).

The following paragraph continues: 'We can say that the twentieth century … has seen more dramatic political developments in the history of mankind' and mentions communism, Marxism, and the war in Dhofar, and, eventually, the succession of Sultan Qaboos and the transformation of the country. Thus, oil slips away from the historical account, and is instead linked to earlier 'problems'. There is no further mention of the transformative role of oil in Oman's national prosperity, much less the event of its exploration and discovery. There is no drawing together of the national body through the shared benefits of oil wealth.

Perhaps more significantly, histories written by those within the Imamate fold do not focus on the exploration of oil either, though there are certainly more references to it there. The two most important historical texts discussing the Imamate war of the 1950s, *Nahḍat al-'ayān bi-hurrīyat 'Umān* (The renaissance of the notables in the freedom of Oman) and *'Umān: Tārīkh Yatakallim* (Oman: history speaks), were published before the discovery of commercial quantities of oil, but certainly during the sequence of exploration and war. *Nahḍat al-'ayān* was first published in 1960 and *'Umān: Tārīkh Yatakallim* in 1963. In addition, the first text has also had several editions, including the last in 1998. The author of both texts, Mohammad bin Abdullah al-Salimi (a son of Nur al-Din al-Salimi, the spiritual force behind the Imamate revival in 1913), barely mentions oil in the first book, but does in interesting ways in the second. It would be conjecture to suggest why oil is more present when covering the same historical periods in the later text, but it may be relevant that the later text was written after Mohammad bin Abdullah al-Salimi spent significant time in exile in Saudi Arabia, where tensions about oil concessions, exploration teams, as well as oil company towns would have been known to him and where he would have been more immersed in international politics.[6] Indeed, by 1963, Iran's oil nationalization movement and the CIA-organized *coup d'état* that removed the enormously popular Prime Minister Mohammad Mossadegh from power were already ten years old. Moreover, Mohammad bin Abdullah al-Salimi served as an emissary of the Imamate government in exile in Saudi Arabia, travelling to Egypt to meet with Gamal Abdel Nasser, to Iraq to meet with Abdul Kareem Qassim, and to Algeria to meet with President Boumediene.

Nahḍat al-'ayān, which is the more famous of the two books, covers the revival of the Imamate in 1913, the election of a new Imam in 1920, the election of the third Imam in 1954, and the early period of the Jebel Akhdar war. However, the entrance of the IPC, with military escort, into Imamate territory is barely mentioned, and this moment marking the beginning of the war is presented as simply the expansionist vision and 'false' unifying ambitions of the Sultan. The only references to oil in this text are exceptionally muted, mentioned in passing in the opening pages of the book (as the author frames his text in a proto-national and modern form) as an economic resource along with copper and other minerals.

Interestingly, here, Mohammad bin Abdullah al-Salimi uses the term '*zayt*', which is oil as in any oily substance, but not gasoline or petrol, which would be '*naft*' or '*bitrūl*'. *Zaytūn*, for example, is the term for olive oil. Indeed, it appears that his would-be audience, in this latter stage of the twentieth century, is still rather unsure of the economic benefits of this 'oil', since he feels it necessary to explain the uses of this *zayt*: 'it is used for cars and aeroplanes' (al-Salimi 1960: 68), he elucidates. *Zayt* (and *naft*) then disappears from this history, appearing only in instances when the author cites another history of Oman published by the Arabian-American Oil Company (ARAMCO), to which he refers simply as *Shirkat al-Zayt*, the 'oil' company.

In his 1963 book, '*Uman: tārīkh yatakallim*, Muhammad bin Abdullah al-Salimi does not mention *zayt* (or *naft* or *bitrūl*) at all when discussing Oman's economy at the beginning of the book. Oil, as pointed out, had not yet been discovered in commercial quantities, though it is interesting that it appears in the earlier book's discussion of national economy but not this later one. Instead, here, the author lists agriculture and metals or minerals (*ma'adin*). This second book, which was meant to be a primer explaining Omani history and the revolution (*al-thawra*) to other Arabic-speakers with whom he was coming into contact in exile, attempts to place Omani history, geography, and basic religious traditions in relation to the rest of the Arabian Peninsula. The text, like many classic historical texts, is structured primarily as a biographical dictionary with lists, in this case, of the Ibadi Imams and then the al-Bu Said Sultans in Oman and Muscat.

However, unlike the previous book, in the final thirty pages of this 260-page volume, al-Salimi outlines the D'Arcy concession agreement in Oman and then the incursion by British oil companies and military escorts into Imamate territory in 1954. The language here is direct and unembellished. He writes, for example: 'the advancing British soldiers entered the town of Ibri and the Imam was resigned to the fact of its occupation' (al-Salimi 1963: 236). The British presence, al-Salimi is clear, is tied to oil exploration.

It should be noted that beyond these histories and textbooks, the most official of national sites – museums – are also decidedly silent about exploration and discovery in Oman. The Bayt al-Zubayr museum, which has served for many years as Oman's primary national museum, does not mention oil at all, and neither do the other national museums in the nation's capital. Even Oman's Oil and Gas Exhibition Centre emphasizes not the events of exploration or discovery, but the science of oil exploration and refining. This is decidedly different from other states in the Arabian Peninsula, where oil exploration, discovery, and export structure national museums and their presentations of transformation. The Dubai Museum, which is housed in the former village fort, for example, has projected a video in the entrance tunnel providing a form of national time-line, where music accompanies images signifying the Emirate's dramatic transformations. Here, even though oil is hardly the basis of Dubai's national economy, the music speeds up as images of rigs appear on the screen. With the oil rigs, the music suddenly shifts from a traditional *oud* to a buoyant and fast-paced exciting rhythm. No such images of oil rigs, accompanied or not by up-beat music, present themselves in Oman's representations of its national history.

Journal of the Royal Anthropological Institute (N.S.), 147-162
© Royal Anthropological Institute 2016

Oil's future: deferred depletion

As I have explored elsewhere (Limbert 2008; 2010; 2015), future oil depletion and potential supplies (rather than its origins) dominate oil-talk in Oman. People's understandings of this future not only differ, of course, but have also changed over the years. Some Omanis almost nostalgically anticipate conditions returning to those prior to 1970 (and oil discovery), with the simplicity of life believed to have been dominant then. Other Omanis worry about such conditions, where 'modern' schools, hospitals, and roads (never mind telecommunications) were all but non-existent. Yet others, meanwhile, insist that human speculation about the future is contrary to proper Islam. In addition, and especially since a scandal in 2005 over Royal Dutch Shell's projections of proven reserves in Oman shook the oil industry, official oil-talk about Oman's future reserves has shifted, as have people's responses to this talk. Greater scientific accuracy has been accompanied by more abstractions as well as people's claims about secrecy.

Despite the differences and changes, it is difficult to over-emphasize the consistency with which Oman is framed, both internally and outside, around the question of future depletion. The future not only frames accounts of the country as well as official self-representations, it also seeps into everyday life. The expectation of a future end of oil appears as people express concern about the abilities (and willingness) of the youth to be self-sufficient or in comments and actions of individuals who state that they expect the end of oil in their lifetimes or soon after. Whether or not people prepare for this future by saving money is another question, and not surprisingly perhaps, most people do not (just as most Americans do not prepare for drought, hurricanes, or economic turbulence even when they may 'know' such calamities are coming within a twenty-year horizon).

It must be noted that this concern over depletion is coupled with two other significant concerns about Oman's future. First, as suggested earlier, there is the understanding, especially in the interior region, of the future possibility of the re-establishment of an Ibadi Imamate, even if not everyone desires this future and even if it is not discussed publicly or generally with foreigners, who may not be trusted. Second, there is the recognition that the Sultan is mortal and has, since 2014, been understood to be ailing. Furthermore, without a known heir, this future without Sultan Qaboos al-Bu Saidi has become particularly and acutely worrisome for many. In light of these two popularly recognized political uncertainties, the question about oil's future and economic stability becomes even more pressing.

The question here, however, is how oil becomes entangled in these religio-political expectations or potentials, and how depletion is understood to cause future changes. That is, how does oil disappear as a cause of economic transformation in the past (and where the 'event' of discovery is rarely publicly described), while remaining a primary concern in the future? And, though, certainly, future depletion is not an event in the same way as a past event would be (since past events are usually more temporally defined), either with clear beginnings and ends or presenting themselves as sudden, depletion remains a constant frame of an expected future transformation. Indeed, future depletion appears publicly in popular discussions and official publications about declining oil production levels. Though the possibilities of natural gas exploration and the improving drilling technologies enabling horizontal and deeper drilling have dampened economic concerns and are said to expand Oman's available reserves, it would be hard to ignore the popular expectation that the future will be decidedly different from the present.

Journal of the Royal Anthropological Institute (N.S.), 147-162
© Royal Anthropological Institute 2016

It is here, perhaps, that we can understand how discovery is a non-event in Oman, while future depletion is an event, and how attention to causation and responsibility highlights the interpretative distinctions between different objects, including natural resources, as well as the ways the same resource may be understood to have the power to (not) effect change. And, further, we can understand better how a substance – in this case, oil – can simultaneously be framed as *not* having effects (or having few effects) in the past, while causing transformation in the future. Whereas the past of discovery is displaced by the arrival of the *Nahda* and the Sultan's rule as the defining event of the country's modernity, the future of depletion encompasses the Sultan's demise as well as the potential for the return of the Imamate, intensifying depletion's transformative effects. This is not to say, of course, that these other happenings will be simultaneous or that the Imamate *will* be re-established. However, the potential for all three to correspond makes this future intensely more foreboding, and very much Oman's potential future event.

Time of plenty, time of drought

Whereas oil is understated and passive in the historical descriptions of the causes and event of its discovery, its depletion – however deferred – is regularly present in everyday talk as well as contemporary accounts of Omani political and economic history. Similarly, the language and frame of depletion, as illustrated by Kamal Sultan's above-mentioned article, is more dramatic, appearing as a future spectre. In contrast to the ways that oil appears and disappears in Omani temporal imaginings and historiography, water has long provided a different kind of politico-temporal imagining associated with its discovery, its disappearance, and its sudden and unexpected reappearance. Here I draw from local histories and conversations to provide examples of the ways water becomes part of and embedded within historical events.

A late twentieth-century local history and biography by a Bahlawi town elder of Sheikh Abu Zayd al-Riyami, the Sheikh who served as the governor of Bahla for thirty years in the first half of the twentieth century as the representative of two Imamate administrations (Imam Salim bin Rashid al-Kharusi and Imam Mohammad bin Abdullah al-Khalili), describes the great drought – *al-mahl* – that afflicted the region in the late 1910s. The author writes:

> In the first years of Abu Zayd's reign, the river of Bahla dried and the wells dried and Bahla closed up from fear and hunger and lack of fruit. That is what people earned for their deeds, and this was from the wisdom and justice of God.

The local history marks the withholding of rain as punishment. God was punishing Bahla because many of its citizens had fought on the side of its local ruler, Nasser bin Humayd, against the establishment of a righteous theocracy. The supporters of the theocracy eventually overtook Bahla in 1916, and the theocratic leadership was able to install its own governor, the locally famous Abu Zayd. Abu Zayd is celebrated in this text for bringing order and justice to the town and, especially, for managing the drought efficiently and effectively, deeds that are later rewarded with rain. The rains mark the solidification of a positive regime. The text continues to celebrate the work of Abu Zayd, who is praised not only for having managed the drought exceptionally successfully, but also for the diligence and hard work with which he organized the labour to restore the town's crumbling canal system.

Similarly, Mohammad bin Abdullah al-Salimi barely mentions oil in his history of the Imamate revival in the first half of the twentieth century, and when he does, he does so

Journal of the Royal Anthropological Institute (N.S.), 147-162
© Royal Anthropological Institute 2016

in very direct language. In contrast, when describing the rule of Bahla before Abu Zayd, al-Salimi's language is entirely different. Here, he describes the opposition to Imamate rule in Bahla under Nasser bin Humayd (whose tyranny is regularly contrasted with the piety of Abu Zayd) and how the pious Imamate forces attack Nasser bin Humayd's army in Bahla to establish righteous rule. The final battle is not only glorious; it is miraculous and accompanied by rain. It is described as follows:

> On the 13th of Rajab 1334 [16 May 1916] the squadrons of Sheikh Rais Hamyar [marched] towards Bahla, and fog clouds concealed them on the road ... The helpers [i.e. the soldiers] left from al-Mahmood [a village near Bahla] after dark on that day under the sudden rain ..., and the victory happened as the thunder urged them on, and the lightning guided them, and they prepared an amazing attack, clashing with what confronted them and storming the wall of Bahla, even though they came to know carnage for it. And from the miracles that adorn history, this lightness of the rain on the attacking freedom fighters gave them the strength to storm the walls of the town (al-Salimi 1960: 282-3).

Rain appears here as a miracle that gives strength and urges the righteous to face death and carnage, and to fight – successfully – for the establishment of righteous rule. Rain (and water in general) not only accompanies events in history, but is itself a powerful actor in shaping its outcome, though, it must be emphasized, it is never a cause independent of God's power. While rain may motivate and encourage, it is, of course, never the first cause.

One might argue that these are older histories of Oman (or local ones of an older tradition) that more readily incorporate divine presence or evidence of the divine in their accounts. That is, one could argue that water carries with it a frame of divine presence because, perhaps, its narrativization is part of a form of history-writing that is structured as awe and amazement at the power of the divine. This is certainly accurate and even more striking when one thinks about the degree to which oil is markedly different.

Here, we should also recognize that we are speaking not simply of 'water', but of rain – a particular form of water: one that humans cannot control (though, of course, this does not mean that humans do not affect it through climate change). Nevertheless, although rain is believed to be beyond human intervention, water in general carries pious and ethical connotations, whereby humans are expected to ensure complete, public access to drinking water and not to allow the distribution of irrigation water to be controlled by the elite. While, certainly, and as I have described elsewhere (Limbert 2001; 2010), such notions and practices are not without their shifting tensions and discussions, the management of water distribution in towns like Bahla has long involved debates and expectations that are considered to be part of pious behaviour. It should be noted, however, that the management of water distribution has also changed dramatically since the 2000s as there has been greater dependence on tankers trucking water from desalination plants on the coast as well as plans to build an enormous pipeline carrying water from the coastal desalination plants to the interior.

Despite the possibly disappearing frame of water in general as necessarily miraculous and evidence of the divine, rain continues to be evidence of the miraculous and of good fortune. The awe expressed when it rains, and when there are storms, gives recognition of God's power and blessing. On a visit to Oman that I made in December 2012, an oil engineer and his wife kindly pointed out that it rained during three of my more recent visits to Bahla in June 2010, in December 2011, and December 2012. 'It is great that you came. Look', the engineer said and paused, 'it rained on each of your visits'. 'Subhān Allah' (praise the Lord), his wife added. Such statements were not simply aimed at flattery or even to acknowledge the closeness of our relationship, but sought to bring another human into the fold of the miraculous and divine action. Though a connection between my presence and rain certainly was not an event in itself, the possible cause of rain – divine blessing – was clear, and the generosity of acknowledging a possible connection was even more evident.

While rain has provided evidence of divine presence, power, or anger and has been seen to cause change, accounts of water discoveries are similarly revealing. Or, rather, the care with which local scholars have held and guarded manuscripts of water histories (such as in Figure 1) suggests the delicacy and power associated with water discovery. Though, as mentioned, trucks, tankers, and pipes from coastal desalination plants have quickly become part of the infrastructure of everyday life in towns in interior Oman, accounts of locally discovered water have been carefully guarded and preserved in secrecy and with diligence. Water management (both record-keeping and practical distrubution) in interior towns such as Bahla had been, until the mid-2000s, divided between different individuals. One man, known as the 'arīf (or knower) controlled the auctioning of water-time and organized the distribution of canal water, while another man held the documents of ownership. This division, it should be noted, was also one

Figure 1. Leather-bound, privately held water documents. (Photo by the author.)

Journal of the Royal Anthropological Institute (N.S.), 147-162
© Royal Anthropological Institute 2016

of status, whereby the former has tended to be lower status, as suggested above, while the latter has been upper status. It should also be noted that until the widespread use of piped water and meters for irrigation in the late 1990s, water ownership for irrigation was determined by time rather than quantity. Individuals and mosques would own, inherit, rent, buy, and sell time, whether thirty seconds or two hours, and the water would be distributed in two-week cycles alternating between day and night.[7]

Documents in Bahla about water-time ownership, I discovered when trying to learn more about water management, were held in very carefully bound manuscripts, locked in briefcases (when available), wrapped gently in cloth, and strictly guarded by elders and their watchful sons. In addition, such documents were not simply lists of which individuals or what mosque, for example, owned how much water-time, though such lists are part of the documentary stash. Rather, the lists were framed by the history of water search and discovery itself, tightly guarded, housed with an upper-status Bahlawi, and maintained in careful and delicate casing. And, as recounted in the opening pages of the carefully guarded book shown in Figure 1, water for the primary Bahlawi canal, *Falaj Mayta*, was discovered by someone with spiritual blessing, Musa bin Ali al-Barakah (a name which itself means blessing).[8] In fact, according to the manuscript account, this particular canal 'continued to flow and even increase during the drought that caused other canals and wells to dry'. The story continues, 'Some people asked how it was possible that these other canals decreased, while this one increased. One person said that we should not ask why'.

Conclusions

In Oman, both oil and water are 'natural' sources of prosperity, either when converted to capital or when used to grow crops that are then sold for profit. Water, however, sustains life, both as drinking water and through the crops it yields for food and subsistence; it also carries with it a fundamental relationship with divine power. The differences between oil and water as natural resources in Oman, however, are apparent not only when examining their uses (and their use and exchange values), but also in the ways they are embedded in the understandings of national and local transformation. To reduce the ethics associated with these natural resources to their basic use not only ignores the ways oil, for example, is tied so intricately to national transformations or national sentiment in many other states, but also limits our understandings of the ways that causality and responsibility are hardly neutral or universally equal.

Oil has enabled the current Omani regime to provide the infrastructure and benefits of schools, hospitals, roads, and government jobs that the previous regime did (and could) not. And yet oil is decidedly absent from the narratives of Oman's modernity. This is an absence that is especially striking given the ways that oil is steeped in the nationalist accounts of other petro-states. Instead of serving as the event that enables the establishment of the modern era or being glorified through stories about the excitement and surprise of discovery, the Omani event of oil appears in its future depletion, generally as a known deferred twenty-year future that might coincide with the end of the current regime (or, at least, the demise of its revered leader) and the possible re-establishment of a more explicit theocracy. Water, in the form of rain in particular, on the other hand, has long marked divine retribution as well as miraculous encouragement and blessing. Such events of divine encouragement and retribution are also evident in the language and moments of water discovery, held, sometimes, as carefully guarded manuscripts and enfolding blessings.

Journal of the Royal Anthropological Institute (N.S.), 147-162
© Royal Anthropological Institute 2016

A severe drought afflicted Oman from 2005 to 2010 and thousands of date palms in Bahla died. Fields of date palms that had been abundant in 2000 had all but completely dried. Canals no longer had water flowing in them. Instead, they held tubes and pipes put down to carry water from the near-dry riverbed and from illegal wells. Illegal pumps also sprang up along the riverbed, taking water to people's fields, and serious disputes between neighbours erupted. Word of the tensions spread to Muscat. Indeed, the future of Oman's fresh water supply is also of grave concern, with heated debates in the country about the relative merits and problems of desalination. Nevertheless, the differences between oil's presence in Omani accounts about national transformation (as a non-event, as unembellished and passive, or as a foreboding and deferred uncertainty) and the ways that water has, historically in any case, appeared (as an actor, as evidence, or as enfolded in a blessing), raises questions about how the drought would have figured in the narration of any significant political transformation. Indeed, though oil's future depletion may be projected, however vaguely, with a time beyond that of the current renaissance and rule of the current Sultan, it would hardly be a surprise if the miraculous presence or absence of rain becomes integral to that future narration as well.

NOTES

This paper was first presented at the American Anthropological Association meetings in 2012 as part of a panel, 'The Eventedness of Nature', organized by Naveeda Khan and Deborah Poole. I also wish to thank Karen Strassler and Laura Kunreuther for their comments, insights, and questions.

[1] It should be noted that placed adjacent to the article by Kamal Sultan (who had suffered a severe stroke in 2000 and has since remained an honorary chairman of his family's company, W.J. Towell) was a commentary by his brother, Shawqi Sultan. The commentary explained Kamal Sultan's love for Oman's landscape and his concern that in the country's speed to provide infrastructure for tourism, the Omani government might destroy the very landscape it aimed to offer. It is unclear why the article was reprinted in 2010 or why Shawqi Sultan felt it necessary to provide a defence of it. Perhaps Shawqi Sultan was feeling similarly, but also aware of a political climate where open critique might signal disloyalty – a particularly poignant accusation for a prominent, yet minority, Shi'a family.

[2] See Fernando Coronil (1997) for an exemplary account of the way oil shapes nationalist sentiment in Venezuela.

[3] Indeed, this paper also emphasizes how some events are not only retroactive, but also expected or anticipated. Gilles Deleuze, according to Alain Badiou, argued that an event may be 'that which has just happened and that which is about to happen, but not that which is happening' (Badiou 2007: 38). While one may disagree with the claim that an event cannot be that which is happening (as many participants in ceremonies or hoped-for revolutionary moments, for example, claim to participate precisely because they want to be witnesses to and *in* the historical event), my point is to emphasize the expectations and anticipatory possibility of events, and not only ones that are 'about to happen'. Messianic as well as apocalyptic expectations would serve as models for such thinking, though here I am interested in a presumably more mundane expectation and anticipation: oil depletion. While, certainly, future depletion – because it is gradual – may not be considered an event in the same way discovery might be understood to have been, the consistent public focus in Oman on twenty more years of oil has also helped shape oil's disappearance as a pivotal future time in Omani history, with a clear 'before' and 'after'.

[4] For a look at fascinating discussions about concessions within Imamate territory, see India Office Records (IOR) R/15/6/24. For British debates about Oman's borders, with reference to oil exploration, see IOR R/15/6/185 and 186. See also IOR R/15/1/625 for another fascinating file on early concessions agreements and discussions in Oman.

[5] See Ennami (1972), Gaiser (2010), Hoffman (2011), and Wilkinson (1987) for excellent English-language publications about Ibadi political philosophy, history, and theology.

[6] For an account of tensions in Saudi Arabia over the Arabian-American Oil Company (ARAMCO) in the 1950s and 1960s, see Vitalis (2009).

[7] For more information and description of the organization and distribution of water in Oman, see Limbert (2010) and Wilkinson (1977). For a description of hydrological terms, see Kharusi and Salman (2015), and for an analysis of the hydraulics of the canal system, see Costa (1983).

[8] Musa bin Ali al-Barakah should not be confused, with the great tenth-century Bahlawi scholar Abu Muhammad Abdullah bin Muhammad al-Bahlawi, otherwise known as the famous scholar Ibn Barakah. I do not know how the two men may have been related, though presumably they were from the same extended family, with the more famous scholar Ibn Barakah being an ancestor.

REFERENCES

BADIOU, A. 2001. *Ethics: an essay on the understanding of evil.* New York: Verso.
——— 2007. The event in Deleuze. *Parrhesia* **2**, 37-44.
BERLANT, L. 2008. Thinking about feeling historical. *Emotion, Space and Society* **1**, 4-9.
CORONIL, F. 1997. *The magical state: nature, money, and modernity in Venezuela.* Chicago: University Press.
COSTA, P. 1983. Notes on traditional hydraulics and agriculture in Oman. *World Archaeology* **14**, 273-95.
ENNAMI, A.K. 1972. *Studies in Ibadism.* Benghazi: University of Libya.
FERGUSON, J. 1999. *Expectations of modernity: myths and meanings of urban life on the Zambian Copperbelt.* Berkeley: University of California Press.
GAISER, A. 2010. *Muslims, scholars, soldiers: the origin and elaboration of the Ibadi Imamate tradition.* Oxford: University Press.
HOFFMAN, V. 2011. *The essentials of Ibadi Islam.* Syracuse: University Press.
HUMPHREY, C. 2008. Reassembling individual subjects: events and decisions in troubled times. *Anthropological Theory* **8**, 357-80.
KAPFERER, B. 2010. In the event – toward an anthropology of generic moments. *Social Analysis* **54**: 3, 1-27.
KHARUSI, N.S. & A. SALMAN 2015. In search of water: hydrological terms in Oman's toponyms. *American Name Society* **63**, 16-29.
LAIDLAW, J. 2010. Agency and responsibility: perhaps you can have too much of a good thing. In *Ordinary ethics: anthropology, language, and action* (ed.) M. Lambek, 143-64. New York: Fordham University Press.
LIMBERT, M. 2001. The senses of water in an Omani town. *Social Text* **19**: 3, 35-55.
——— 2008. Depleted futures: anticipating the end of oil in Oman. In *Timely assets: the politics of resources and their temporalities* (eds) E. Ferry & M. Limbert, 25-50. Santa Fe, N.M.: School of Advanced Research Press.
——— 2010. *In the time of oil: piety, memory, and social life in an Omani town.* Palo Alto, Calif.: Stanford University Press.
——— 2015. Reserves, secrecy, and the science of oil prognostication in Southern Arabia. In *Subterranean estates: life worlds of oil and gas* (eds) H. Appel, A. Mason & M. Watts, 340-52. Ithaca, N.Y.: Cornell University Press.
MINISTRY OF EDUCATION 2007. *Al-tarbīyat al-mūwāṭinīya fī fikr al-Sulṭān Qābūs* [National education in the thought of Sultan Qaboos]. Beirut.
MINISTRY OF OIL AND GAS n.d. Brief history of oil and gas in Sultanate of Oman (available on-line: *http://www.mog.gov.om/Portals/1/pdf/oil/history-Oil-Gas-en.pdf*, accessed 12 January 2016).
PETERSON, J.E. 2007. *Oman's insurgencies: the Sultanate's struggle for supremacy.* London: Saqi Press.
AL-SALIMI, M.B.A. 1960. *Nahḍat al-ʿayān bi-hurīyat ʿUmān* [The renaissance of the notables in the freedom of Oman]. Cairo.
——— 1963. *ʿUmān: tārīkh yatakallim* [Oman: history speaks]. Damascus.
SULTAN, K. 2010. When our oil is exhausted [with commentary by S. Sultan]. *The Muscat Daily,* 8 June.
VITALIS, R. 2009. *America's kingdom: mythmaking on the Saudi oil frontier.* New York: Verso.
WILKINSON, J.C. 1977. *Water and tribal settlement in South East Arabia: a study of the Aflāj of Oman.* Oxford: University Press.
——— 1987. *The Imamate tradition of Oman.* Cambridge: University Press.

Oman liquide : pétrole, eau et causalité dans le sud de l'Arabie

Résumé

Le présent article explore la manière dont différentes ressources naturelles apparaissent dans les imaginaires temporels. La question posée par l'auteure est : comment le pétrole et l'eau en viennent-ils à encadrer les relations et les chronologies de la transformation et, plus particulièrement, de la causalité ? Pour comprendre les visions des futurs environnementaux, il nous faut non seulement examiner les formes de planification, d'attente et de pronostic qui modèlent les connaissances ou les perceptions du futur, mais aussi nous demander comment et pourquoi des liens de causalité et des événements significatifs sont

Journal of the Royal Anthropological Institute (N.S.), 147-162
© Royal Anthropological Institute 2016

associés à des ressources naturelles données. Sur la base de son travail antérieur explorant l'orientation future du discours sur l'épuisement du pétrole à Oman ainsi que de sources textuelles et de l'ethnographie, elle avance qu'alors que l'eau est associée à la piété et à la présence divine, le pétrole a été considéré comme beaucoup plus transitoire et comme le produit des interventions et politiques humaines, qui se manifestent dans un contexte historique et politique chargé. Alors que l'eau semble motiver les événements et causer les changements, servant d'indication et de véhicule du pouvoir de Dieu, le pétrole apparaît moins comme une cause de transformation nationale, du moins à ses débuts.

9

Prognosis past: the temporal politics of disaster in Colombia

AUSTIN ZEIDERMAN *London School of Economics and Political Science*

In this paper, I explore a prognostic modality of environmental politics I call *risk in retrospect*. I do so by examining conflicts that are not just over the ability of state science to know and govern the future, but also its failure to have been able to do so in the past. I begin by discussing the 2009 earthquake that hit the Italian city of L'Aquila and then turn to a similar case in Colombia, both of which reflect the constitutive relationship between political authority and foresight. I then trace the strong historical precedent for this type of political situation in Colombia by discussing a key cultural referent – Gabriel García Márquez's 1981 book, *Chronicle of a death foretold* – as well as the convergence of two catastrophic events during one week in November 1985 that were both seen, in their aftermath, as 'tragedies foretold'. I conclude by considering what it would mean for prognosis to become the terrain on which citizens engage in political relationships with the state.

This paper about Colombia begins in Italy, for Italy is home to a recent example of the prognostic modality of environmental politics. The situation I refer to is the 2009 earthquake that hit the Italian city of L'Aquila and the judicial decision handed down in 2012 to convict six scientists and one government official of manslaughter. They were guilty, the court ruled, of failing to give adequate warning to residents about the risk of an impending disaster – or, to paraphrase some of the more sensationalist media headlines, of not being able to predict the future. This case is significant not only because it may have far-reaching consequences for the politics of disaster risk management or because it raises so many questions about the relationship between technical expertise, political authority, and legal accountability. It is also worth mentioning here for its ability to highlight the peculiar temporality of conflicts over the state's responsibility to know and govern the future, and its failure to have been able to do so in the past. In other words, it reflects a prognostic modality of environmental politics we might call *risk in retrospect.*

In early 2009, a series of small tremors was felt across the Abruzzo region east of Rome. The public's unease was heightened when a retired local laboratory technician began issuing unofficial predictions, based on homemade monitoring devices, that a major earthquake was imminent. The national risk commission then met in the city of L'Aquila on 31 March to evaluate the situation, concluding that the recent swarm of seismic activity only slightly increased the probability of a major event. In a press conference, a government spokesman communicated that there was no danger and that it was unnecessary to issue safety warnings or order evacuations. A week later, a 6.3 magnitude quake devastated the city, killing 309 people. Seven members of the risk commission were eventually taken to court by lawyers representing the victims' families. In support of the seismologists and geologists standing trial, an international coalition of more than 5,000 scientists issued an open letter to President Giorgio Napolitano. Their letter highlighted the scientific consensus that seismologists cannot predict when and where an earthquake will occur, and argued that the defendants were being prosecuted for failing to do the impossible. This was the primary argument put forth by the defence. But on 22 October 2012, the prosecution prevailed. Judge Marco Billi ruled that the defendants had communicated 'inexact, incomplete, and contradictory' information about the risk of a major earthquake; that they had lulled the public into a false sense of security with a deceptively reassuring statement; and that they had failed to give clear preparedness advice. The verdict: all seven were sentenced to six years in prison and ordered to pay over $10 million for court costs and damages.

Around the time of this ruling, Colombia was suffering from one of the worst rainy seasons on record. Amidst the deluge, President Juan Manuel Santos presided over the twenty-fifth anniversary of another catastrophic event: a volcanic eruption in 1985 that set off massive mudslides and buried the town of Armero, killing nearly 30,000 people. In his commemorative address, he urged Colombians to apply lessons learned from the earlier disaster to the current one in order to avoid falling into 'victim syndrome' (*síndrome del damnificado*).[1] Santos encouraged those affected by the recent storms to adopt 'an attitude of mutual collaboration and solidarity' and to emulate the patience and perseverance of the 13-year-old girl, Omayra Sánchez, whose tragic death had come to symbolize the calamity. Moreover, he signalled that 'among the lessons this great tragedy left behind is the importance of foresight and doing everything possible to prevent tragedies'. In the case of Armero, he recalled, 'it may have been possible to avoid many of the deaths ... this is an important lesson ... if only the warnings had been heard'. Santos reminded his audience that the 1985 event had given birth to Colombia's national system of disaster prevention, which had since saved thousands of lives. His comments demonstrated the degree to which the Armero tragedy continues to underpin the state's political, legal, and ethical responsibility to anticipate and prevent potentially catastrophic events.

The Italian and Colombian cases both sit within the domain of *prognostic politics*, the topic of this special issue, which draws our attention to the calculative techniques used to predict environmental futures, and to the political significance of such predictions. However, this paper focuses on natural *disasters* rather than natural *resources* – social *bads* rather than social *goods*. Yet the temporal dynamic it highlights might be as common to debates over declining water supplies as to those regarding rising sea levels. For attempts to grasp both resource availability *and* environmental risk rely on techniques for calculating the likelihood of different futures, and then bringing them into the realm of individual or collective decision-making. Despite a wide range of techniques of prediction and degrees of predictability, both resources *and* disasters are situated in a common technocratic domain, within which political authority is linked to the state's responsibility to generate knowledge about the future and to govern accordingly.[2] As Reinhart Koselleck (2004) has argued, (rational) prognosis displaced (religious) prophecy as the paradigmatic conception of futurity in the modern period along with the secularization of political authority. That said, the L'Aquila disaster in Italy and the Armero tragedy in Colombia suggest that prognostic politics is not solely about the future, but often also about the past. Both cases illustrate a peculiar temporality – at once prospective and retrospective – that often characterizes public controversies in the aftermath of catastrophe.

The following analysis builds on long-term fieldwork in Colombia to illuminate a prognostic modality of environmental politics: *risk in retrospect*. Over a twenty-month period from August 2008 to April 2010, I conducted both ethnographic and archival research in Bogotá on the emergence of disaster risk management as a technique of urban planning and government (Zeiderman 2012; 2013). My specific focus was a municipal housing programme actively relocating people from what in the early 2000s had been designated *zonas de alto riesgo*, or 'zones of high risk'. These were areas on the periphery of the city deemed vulnerable to landslides and, in some cases, also to floods. I returned for a one-month follow-up visit in January 2012, towards the end of what would eventually amount to over a year of inordinately heavy rainfall, to find areas near my fieldsites recovering from extensive inundation. I was interested in the government's response to the flooding as well as that of the people whose homes had been damaged by it, and made a series of visits to the affected areas. These neighbourhoods were outside my initial study area, so I had to rely on chance encounters with local residents who responded to my inquiries, invited me into their homes, and agreed to be interviewed. In order to reconstruct the events leading up to the floods, I supplemented interview data by consulting media coverage from December 2010 to February 2012.[3]

In what follows, I draw on this material to examine conflicts over the ability of state science to know and govern environmental futures and the failure to have done so in the past. I begin by considering differing approaches to the politics of temporality, especially in relation to a paradigmatic spatial configuration – the city – in which prognosis has become a key idiom of political authority and responsibility. Having introduced the problematic of *risk in retrospect* by way of the 2009 earthquake that hit the Italian city of L'Aquila, I then turn to a more fine-grained analysis of a similar case in Colombia that transpired during and after the worst rainy season in recorded history. I trace the historical precedent for this type of situation in Colombia by discussing a key cultural referent – Gabriel García Márquez's 1981 book, *Chronicle of a death foretold* – as well as the convergence of two catastrophic events during one week in November 1985 that were both seen, in their aftermath, as 'tragedies foretold'. I conclude by considering what it would mean for prognosis to become the terrain on which citizens, especially the poor, engage in political relationships with the state. Ultimately, I argue that if political authority is increasingly tied to the ability to generate knowledge about the future – of resources, of disasters, and of many other social and environmental goods and bads – our capacity to understand the strategic potential of prognosis is of pre-eminent importance.

The politics of temporality in the city

The L'Aquila disaster in Italy and the Armero tragedy in Colombia both point to a peculiar yet widespread modality of environmental politics that is at once prospective and retrospective. This observation is inspired by historian of science and technology Paul Edwards (2012), who has argued that the extent of two recent disasters – the Fukushima nuclear fallout and the Deepwater Horizon oil spill – could not be known except through already-existing computer simulations that could generate predictions of what was likely to happen in these sorts of events. That is, since tracking the real-time spread of the actual disasters was impossible, models designed previously to simulate the potential release of toxic substances (radiation and oil) into the surrounding environment became the authoritative record of what was happening (cf. Edwards 2012). In other work, Edwards (2010) has shown how the science of global warming also relies on computer models and simulations, and that without them there are no data with which to predict how the world's climate is likely to change. In the case of the two recent disasters, however, he complicates the temporality of models and simulations by showing how they enable scientists not only to foresee what will happen, but also to assess what has already happened. This temporal reversal – seeing prediction as about both the future and the past – is crucial to what I am emphasizing here.[4]

That disasters are often understood through non-linear temporalities should come as no surprise to anthropologists, given the discipline's long-standing interest in time (Bear 2014; Geertz 1973; Gell 1996; Greenhouse 1996; Munn 1992; Rabinow 2008). What may be more novel is the observation that temporality is explicitly political, such that the state's authority is often predicated on specific arrangements of time. Referring to the 'politics of temporality' in the contemporary United States, Vincanne Adams, Michelle Murphy, and Adele Clarke argue that future-orientated regimes of *anticipation* increasingly define how 'we think about, feel and address our contemporary problems' (2009: 248). In their view, 'anticipation is intensifying into a hegemonic formation' that is motivating speculative logics of capital accumulation, spreading through our institutions of government, and becoming an affective state that orientates individual and collective behaviour. The sweeping phenomenon these scholars describe resonates throughout other works of social theory, which highlight the political implications of heightened anxiety about futurity and the subsequent proliferation of pre-emptive actions (Cooper 2006; 2010; Martin 2007; Massumi 2009; 2010).

While much can be gained by theorizing temporal transformations on a grand scale, or what Daniel Rosenberg and Susan Harding refer to as the 'big stories of the future' (2005: 14), they often lead to reductive denunciations of temporality as a domain of hegemony and domination.[5] Once we acknowledge Koselleck's irrefutable point that historical time 'is bound up with social and political actions, with concretely acting and suffering human beings and their institutions and organizations' (2004: 2; cf. 2002), then it logically follows that all forms of temporality are suffused with and constitutive of power (Greenhouse 1996). Yet while this ought to lead us to analyse futurity and its uneven effects on certain bodies, spaces, and populations, it would be a mistake to resort to automatic denunciations. For doing so would risk ascribing an almost totalizing quality to the politics of time, and leave us devoid of resources for thinking about the heterodox temporalities existing alongside and within dominant or hegemonic formations.[6] We need tools for analysing futurity as constitutive of both political authority and political possibility.[7]

AbdouMaliq Simone's (2010) understanding of 'anticipatory urban politics' is helpful in this regard.[8] Among the urban poor in Northern Jakarta, Simone finds people 'reading the anticipated manoeuvres of stronger actors and forces and assessing where there might be a useful opportunity to become an obstacle or facilitator for the aspirations of others' (2010: 96).[9] But the politics of anticipation is 'not just a form of resistance or simply a politics from below'.[10] Anticipatory politics constitutes a future-orientated 'game of transactions' that brings differently positioned urban actors into contact with each other, resulting in benefits and constraints for all (2010: 101). The future is neither the exclusive domain of state planners and economic elites nor that of popular political mobilizations – it is a horizon of strategic politics for diverse attempts to make and remake the city in the present. Since the city is 'not hinged, not anchored' to any single trajectory, Simone urges us to remember that 'by definition [it] goes toward many different futures at once' (2010: 115).

Simone's reflections suggest that temporality is mapped on to certain spatial configurations, such that the city, we might say, is a key 'chronotope of modern time' (Bakhtin 1981; Bear 2014). In cities, the future is alongside us, upon us, all around us. The past, of course, is present, too.[11] But it is tempting to assume that (in some fundamental way) the city is orientated temporally towards horizons of possibility, expectations of change, and the anticipation of things to come. There is something misleading about this assumption, however, since it is rather peculiar to cities belonging to the category of the 'modern'. One of the defining characteristics of modernity was a belief in the progression of time towards a more efficient, more prosperous, and all-around better future; in turn, this promise hinged on the growth and development of cities and the domination of nature.[12] The 'modern city' was considered the most advanced stage of social evolution, historical progress, and cultural development against which all other spatial forms were judged. And the destiny of places outside of Europe and North America was presumed to be a perpetual game of catch-up with Paris, London, and New York (Robinson 2004; Roy 2008).

These imagined futures came into question in the late twentieth century as teleological, evolutionist, and developmental narratives of all kinds began to lose credibility (Ferguson 2006: 176-93). Today, it is more common to foresee futures of crisis, chaos, and catastrophe than to envision the emergence of ideal spaces and societies – conceptual resources for imagining alternative possibilities appear to be in short supply. Susan Buck-Morss (2000) famously attributed the 'passing of mass utopia', as she called it, to the exhaustion of the collective dreams of historical progress previously shared by East and West during the Cold War. With phrases like the 'end of history' (Fukuyama 1992), the 'death of progress' (Levitas 1982), the 'post-development era' (Escobar 1992), and 'living in the end times' (Žižek 2010), others have made similarly epochal claims about the demise of the utopian imagination or of progressive thinking more broadly. These proclamations no doubt over-reach, and yet the weakening of grand narratives like 'modernization' and 'development' has made way for a radically different future – one neither certain nor desirable, but rather filled with uncertainty and marked by a rather grim sense of how history will unfold. Disaster, even apocalypse, looms large on the horizon.[13]

The notion that we are now living 'in the wake of utopia', geographer David Pinder (2005: 12) points out, has greatly influenced dominant thinking about urban environments. In this context, inherited frameworks for urban planning and governance have been undergoing significant transformations as concepts like 'preparedness', 'resilience', and 'future-proofing' have taken centre stage. Contemporary visions of urban environmental futurity share the sense that the city is a space of menacing uncertainty, imminent threat, and potential collapse. The logical response is the rationality of security underpinned by techniques for managing risks – of natural disaster and resource scarcity, but also of financial crisis, political violence, and disease outbreak. These techniques are multiple, as they have been assembled in order to govern different forms and degrees of uncertainty attached to diverse objects and events – predicting the likelihood of an earthquake can be altogether different from calculating the probability of a terrorist attack. Even when dealing with the same type of object or event, different actors (e.g. fluvial geomorphologists, people living in a floodplain, insurance companies) employ different forms of prediction that enable different sorts of claims and sets of responsibilities.

As conventional assumptions about futurity are being thrown into doubt, we see something akin to what Rosenberg and Harding call the 'crisis of modern futurity' (2005*a*: 4), here in a specifically urban environmental form. Novel regimes of futurity are emerging in cities throughout the Global North and South, which generate new forms of political, legal, and ethical responsibility tied to the imperative to predict and mitigate future catastrophes (cf. Laidlaw 2013).[14] Prognosis then becomes a terrain on which citizens engage in political relationships with the state. This was true of the 2010-11 flooding emergencies in Colombia, as well as of the 2009 earthquake in Italy. But while both events involved the state's failure to foresee environmental futures, there is a difference between looking *forward* to potential disasters looming on the horizon and looking *backward* to ones that have already occurred. As we will see, *risk in retrospect* is a prognostic modality of environmental politics that sustains both political authority and critique.

After the flood

From late 2010 to early 2012, Colombia was besieged by the worst rains on record. Unusually high levels of precipitation were attributed both to La Niña, or the cyclical cooling of the eastern Pacific Ocean that disrupts typical weather patterns, and to the increased severity of meteorological events associated with climate change. The resulting floods and landslides displaced hundreds of thousands of people, destroyed homes throughout the country, severely impacted the economy, and took the lives of several hundred citizens. Although rural areas in the north of the country and along the Caribbean coast were hardest hit, urban areas, such as a suburb of Colombia's second largest city, Medellín, also witnessed grave tragedies. We could approach this situation by examining the models used to generate rainfall predictions and flood risk maps and their role in shaping the distribution of destruction and displacement. In temporal terms, our analytical position would then be both retrospective and prospective: looking back to the past to reveal the state's failure to prevent future disasters. Instead, I would like to draw attention to some of the conflicts that emerged during and after this period of flooding, which themselves revolved around both hindsight and foresight. Focusing on a specific flood event in the capital city, I will show how this temporal orientation was the ground for both political authority and critique.

On Monday, 5 December 2011, Bogotá's largest daily newspaper, *El Tiempo*, reported that, following a forty-eight-hour deluge, the Bogotá River was nearly overflowing its banks. Andrés González, the governor of Cundinamarca, the department that surrounds Bogotá, cautioned that water levels had risen sharply and could soon cause flooding downstream in the capital. These predictions came true the following afternoon. Another rainstorm developed, and the canals draining *aguas negras* (wastewater) from the city were blocked from conveying effluent to the Bogotá River, as they do under normal circumstances, owing to the river's abnormally high level. Instead of discharging storm water and sewage, one of the drainage canals began to function as a reservoir, collecting wastewater and distributing it laterally across residential neighbourhoods in the peripheral locality of Bosa on Bogotá's southwestern edge. Wastewater not only rushed through the streets, but also percolated up through sewers and drains, and within hours had risen to nearly a metre in height. Residents retreated to the second floor of their homes until rescue workers arrived with inflatable boats to transport them to dry land. Interim Mayor Clara López then declared an emergency and called on her staff to identify where to discharge the Bogotá River's excess flow while pleading with *bogotanos* to limit water use immediately.[15] These responses were effective though insufficient, and the municipal water utility appealed to neighbouring countries to lend Bogotá a high-capacity pump used normally for petroleum extraction.[16] Aided by a respite of dry, sunny weather, these measures eventually succeeded in reducing the river's force and draining the inundated neighbourhoods.

Mayor López then accompanied President Santos to El Recreo, one of the most heavily affected areas, to begin the provision of humanitarian aid. López had already agreed to provide tax relief as well as a rental subsidy of 550,000 pesos (US$300) to those forced by the flooding to evacuate their homes.[17] In addition, the municipal water utility had guaranteed fifteen days of free delivery.[18] Donning a red 'Colombia Humanitaria' windbreaker, Santos announced that the national government would provide an additional 1.5 million pesos (US$775) to each family adversely affected by the flooding.[19] The President estimated that more than 10,000 families would be eligible for aid, but said this was contingent upon an official census of the victims.[20] The Secretary of Social Integration began compiling a registry by consulting the administrator of each residential housing complex and other community leaders. Bogotá's disaster-risk-management agency (FOPAE) followed with house-by-house inspections to determine the extent of property damage and the number of people affected. Although preliminary estimates were as low as 5,000 victims, FOPAE soon announced that the flooding had impacted over 45,000 people in total.[21] Once the official census was complete, the government issued a certificate of damages to each household, and the relief promised by López and Santos could be disbursed.[22]

Long lines formed wherever aid was being handed out, and residents declared that there was not enough for the number of families in need. Reports came back that there was no systematic process for allocating assistance; that food rations were being proportioned in an *ad hoc* manner.[23] Meanwhile, rumours circulated of people with false certificates arriving from elsewhere to steal from those in need. Others complained of flaws in the census, claiming they had been wrongfully excluded. For several hours, protesters blocked the Avenida Ciudad de Cali, a main thoroughfare in southwestern Bogotá, citing unreasonable delays in the distribution of aid.[24] Reporters from the news magazine *Semana* visited the flood zone in January and found the neighbourhood rife with accusations that the government had not made good on its promise to disburse subsidies by Christmas.[25] In response to the pressure, the Secretary of Government ordered a review of the census, and FOPAE returned to verify their count. With media reports and popular protests denouncing the government's response as disorganized and exclusionary, draining the city of flooding turned out to be less of a challenge than providing aid to the victims.

The focus of public discontent soon shifted, however, as residents from the flooded neighbourhoods began blaming the authorities for having ignored the fact that the Bogotá River had overflowed its banks three times in recent years. They criticized the state for permitting development in high-risk zones and demanded accountability from the construction company that had knowingly built housing complexes in flood-prone areas.[26] A group of 150 protesters blocked a local transportation hub, Portal de las

Américas, drawing attention to the fact that the flooding had predominantly inundated public housing. A resident framed the issue as follows: 'These houses turned out to be a scam (*el paquete chileno*): [the state] promised us development (*valorización y vías*) and we ended up living in a gutter'.[27] Shared among these critiques was the sense that the state had failed to take the precautions necessary to protect people from a known threat. That is, the government should have foreseen the disaster and, if unable to prevent it, at least done more to mitigate its adverse effects.

As those affected by the flooding were making their voices heard, similar perspectives circulated in the media. In the editorial page of *El Tiempo*, author and columnist Yolanda Reyes cited a history of irresponsible housing policy:

> I remember when Mayor Peñalosa invited 'opinion leaders' on a helicopter tour of the urban developments of Metrovivienda [social housing] in Bosa. Journalists admired the bicycle routes, the avenues, and the public spaces on the edge of what is now a sewer of *aguas negras* ... Against logic, [Metrovivienda] developed these urbanizations with the promise that risk mitigation would happen later.[28]

Reyes also refuted President Santos's constant references to La Niña by blaming the government for not having taken preventive measures and for ignoring past flooding experiences:

> [T]hey know that what happened in Bogotá is not new, that it is not simply the consequence of global climate change or the fault of '*la maldita niña*', but rather of *maldita improvisación* [damned improvisation]. If they consulted historical documents they would see that from Mosquera to San Victorino [a large area of western Bogotá], there have been times when those who were sold on the model homes in Bosa had to rely on the very same inflatable boats that are carrying them away today.

Emphasizing the error of uncontrolled growth of the urban periphery, Reyes asked: 'Which public agencies approved those licences and allowed these developers to profit? Was there any fine print that warned about the probability of flooding?' She concluded that protecting the urban poor from potential disasters is a central responsibility of the state.

As Reyes's op-ed makes clear, flooding in Bogotá was discussed publicly as more than just a problem of housing policy and land-use planning; issues of prognosis were front and centre. In an interview with Mayor López, *El Tiempo* asked explicitly whether the emergency in Bosa could have been prevented. 'It was foreseen', she admitted, 'and this is why we built flood control channels, maintained the drainage systems, and dredged the Bogotá River'.[29] Looking ahead to the future, the media also initiated discussions about how to ensure the same situation would not repeat itself during the next rainy season. In another op-ed, the chief editor of *El Tiempo*, Ernesto Cortés Fierro, argued that more anticipatory governance of 'zones of high risk' would be necessary: '[T]here are a million and a half souls sitting on the bank of the river in permanent danger, and rather than just asking ourselves what they are doing there, we have to respond to the question, what are we going to do about it?'[30] A range of infrastructural solutions were considered, such as removing the sediment and debris that impeded drainage through the system and widening the river to increase its conveyance capacity.[31] When asked whether Bogotá was prepared for more rain, Secretary of Government Antonio Navarro Wolff assured the public that the administration would prioritize additional preventive measures so the same problems would not return in the future.[32]

Journal of the Royal Anthropological Institute (N.S.), 163-180
© Royal Anthropological Institute 2016

Hindsight and foresight

In January 2012, I visited the neighbourhoods that had been inundated by storm water and sewage one month before. I rode the TransMilenio mass transit system from central Bogotá to its southwestern terminus, the Portal de las Américas, which protesters had brought to a standstill in the days after the flooding. As I transferred to a bus that ran alongside one of the canals conveying wastewater to the outskirts of the city, I detected the stench of raw sewage still floating in the air. Debris lay stranded along its edges and hung from the bottom branches of squat trees lining its banks. I got off at a bakery and spoke to a waitress who confirmed that I was getting close to the site of the floods. The water level had risen a couple of feet in this part of the neighbourhood, she said, and the bakery had been forced to close for days. She then indicated how to get to the area that had been badly affected. 'It's pretty much all cleaned up by now', she said, 'but you'll find people who are still angry. They knew something like this was going to happen, but nobody would listen'.

Following her advice, I continued in the direction of the Bogotá River and found a number of residents willing to recount the story of the flooding. They told me about the panic they felt as the waters rose all around them, the hardship of cleaning up and drying out, and the government's dysfunctional response. But running through many of my conversations was also the sense that this was an event that could have been predicted, and therefore prevented. Mobilizing different forms of evidence – personal experience with past flooding events, periodic monitoring of water levels in the Bogotá River and the condition of adjacent drainage canals, the government's own risk assessments and mitigation efforts – residents claimed that the flooding was the result of prognostic failure on the part of the state.

The majority of the buildings I passed on my visit were *vivienda de interés social* (or social interest housing) constructed in the last ten to fifteen years. Although this was the urban periphery in a geographical sense – I could see where the pavement ended and the pasture began – it was no squatter settlement that had sprung up in a flood zone because people had nowhere else to live. These housing complexes had been officially planned and built with public funds. I stopped in a small convenience store operating out of the front room of a two-storey townhouse and began chatting with the woman behind the counter. When I asked about the flooding, she gestured to the point on her leg up to where the water had risen. 'You've done a good job getting everything back into shape', I remarked, noting the absence of visible damage. 'Yeah, but it hasn't been easy. The government promised us three *ayudas* [disbursals of aid], but we have only received one'. She then told me that people in the neighbourhood were planning legal action against the state and the developer. 'If you want to know more', she offered, 'you should talk to Doña Lucia, the president of the housing complex. She runs a *supermercado* on the next block'.

I did as she said, and found Doña Lucia in a shop similar to the one I had just visited, except that it stocked fresh fruits and vegetables and had a small refrigerator case for meat and dairy. At first, Doña Lucia seemed hesitant to talk about the flooding: 'What exactly do you want to know?' 'I'm interested in the government's response to the emergency', I answered, 'and how the people who live here feel about it'. This caught her attention, and she showed me the spot where President Santos addressed the residents of the neighbourhood, promising each victim a subsidy of 1.5 million pesos. 'Ever since then', she said shaking her head,

Journal of the Royal Anthropological Institute (N.S.), 163-180
© Royal Anthropological Institute 2016

there has been major disorganization on the part of the government in fulfilling its promises. Four days after Santos stood right outside my house, they came to conduct a census. This was necessary since there were lots of *gente deshonesta* [dishonest people] trying to pass themselves off as victims. But the census was completely disorganized. Some people were listed under the wrong identification number, others were excluded altogether, and this made it difficult for those who needed help to get it.

El desorden del estado, she called it. 'The disorder of the state'.

'Is that what the protests have been about?', I asked. Doña Lucia responded with an important clarification:

Well, there were some people who tried to falsify their identity in order to get themselves counted by the census. They are not from here. But those demonstrating in the Portal de la Américas were promised subsidies that never arrived. Many of the *tomas* [protests] have been organized by people whose houses had flooded.

I then asked whether these demonstrations had made a difference. 'They came back to conduct another census', she noted, 'so I suppose we forced them to try to get it right. But there are still many people who deserve assistance and have not gotten any help. However, the main problem, as I see it, is that none of this should ever have happened in the first place'.

Surprised by this abrupt change of direction, I asked Doña Lucia what she meant. She gave me the following explanation:

We all knew this was going to happen. A group of us, maybe forty in total, got together back in September to call attention to the risk of flooding in the area and to denounce the government for not doing anything to prevent it. Look it up. It was on 8 September. City TV, Caracol, RCN, all the networks came to cover the story. We told them they had to clean out all the debris in the canals before the next heavy rain. Acueducto [the water utility] did a bit of clean-up, but in general the government just ignored us. And as a result, they were not prepared in December when the weather got really bad. We knew this was going to happen. They did too, but they didn't do anything. They just waited, and then opened the floodgates once it was already too late. They have to be more prepared. The *olas invernales* [winter storms] are just going to get stronger and stronger every year. We're worried! The river is going to fill up again next time there are heavy rains. We're always watching the sky.

After our conversation, I did indeed look the story up, as Doña Lucia had instructed me to do. I started with the archive of *El Tiempo*, but found nothing indicating that residents from Bosa had made a *denuncia* (denunciation) back in September about the risk of flooding in their neighbourhood. I then looked through past editions of the other major Bogotá newspaper, *El Espectador*, and found nothing there either. I checked a few more media sources, but again came up empty. I was about to give up when I came across a posting on the RCN Television website from 9 December 2011 with the heading: 'Flood in Bosa had been denounced three months ago'.[33] Sure enough, what Doña Lucia had said was true. And since I could find no record of the protest until after the flooding had already occurred, it seemed that the warning had indeed fallen on deaf ears.

The posting on the RCN website linked to a video clip that revealed a group of demonstrators gathered in front of television cameras on the morning of 8 September in Bosa (Fig. 1). Some were dressed formally in suit and tie while others were bundled up with scarves and hoods to ward off the cold morning air. Men and women of all ages faced the cameras with expressions conveying seriousness and conviction. The newscaster explained that residents were worried that the area's drainage system was inadequate to withstand the coming winter rains. Aided by handwritten signs, they

Figure 1. Disaster warning, September 8, 2011. (Source: Noticias RCN.)

warned of a potential threat that could lead to grave consequences, and asked the government to take immediate action (Fig. 2).

> *Exigimos que la Alcaldía Distrital solucione el problema del río. ¡Evitemos una gran tragedia!* We demand that City Hall solve the problem of the river. Let's avoid a huge tragedy!

> *El Consejo de Kasay de los Venados II presente... Ante el inminente riesgo de desbordamiento del río no queremos un gobierno distrital indiferente.* The Administrative Council of Kasay de los Venados II [housing complex] is present ... in light of the imminent risk of the river overflowing, we don't want an indifferent city government.

The microphone was passed to a middle-aged man, who took on the responsibility of speaking for the group:

> Our worry is that we are staring in the face of an imminent flood (*ante una inminente inundación*) throughout the neighbourhood of El Recreo, and therefore we are making an urgent call to the Empresa Acueducto [water utility] to please dredge this river ... this drainage canal ... in an urgent manner. And we call not just on Acueducto but also on the municipal government to help us, because as soon as this area floods, all possible solutions are going to be unnecessary ... they won't do us any good at all.

Reflecting the peculiar temporality of environmental politics that I am calling *risk in retrospect*, the video clip then jumped ahead to footage of the December floods, driving home the point that these pleas were not heard and the residents of Bosa were now facing the consequences.

Journal of the Royal Anthropological Institute (N.S.), 163-180
© Royal Anthropological Institute 2016

Figure 2. The imminent risk, September 8, 2011. (Source: Noticias RCN.)

Tragedies foretold

Like the debate surrounding the culpability of the Italian earthquake scientists, the case of flooding in Colombia involved not just the state's responsibility to know and govern the future, but also its failure to have done so in the past. This type of situation has strong precedent in Colombia. A book by Gabriel García Márquez, *Chronicle of a death foretold* (*Crónica de una muerte anunciada*) (1981), is a key cultural referent that has been used repeatedly, in Colombia and abroad, to give expression to tragedies that were foreseen but not avoided. Though not his most celebrated work, its circulation was boosted by the fact that García Márquez was awarded the Nobel Prize in Literature the year after its publication. An allegorical tale based on historical events that took place in the early 1950s in a sleepy town near the Caribbean coast, the novel is frequently referenced in ordinary conversation, but also by the media and politicians, to characterize undesirable occurrences that were anticipated but not averted. While it would be an exaggeration to say that the structure of *Chronicle of a death foretold* typifies a particularly Colombian understanding of temporality, it nevertheless helps explain the prognostic modality of environmental politics under examination here. And while this modality is not uncommon, García Márquez lent it both rhetorical form and moral authority such that it would become readily available as a frame for interpreting subsequent tragic events.

The book tells the story of the murder of Santiago Nasar by the brothers of Ángela Vicario, a woman whose virginity Santiago had supposedly taken. Ángela's husband discovers his bride's lost innocence on the night of their wedding and immediately returns her to her family. Ángela names Santiago as the culprit, and her brothers, butchers by trade, set out with knives to avenge their sister's honour. Intoxicated from the previous night's festivities, the Vicario brothers are not discreet about their objective. They are seen all over town sharpening their weapons, preparing their attack, even boasting about their plan to kill the man who has sullied Ángela's reputation. Nearly everyone knows what is going to happen, and yet no one intervenes to prevent the murder. There are different reasons for this: some townspeople doubt the sincerity of the Vicario brothers' threats; others attempt unsuccessfully to warn the intended victim. In the end, Santiago is stabbed to death on the threshold of his family home.

The central problematic and temporal structure of this short work by the most recognized of modern Colombian authors parallel the prognostic modality of environmental politics described above. The narrator returns to the town many years after the murder, only to find evidence that everyone had known the killers' intentions ahead of time and had been unable to stop them from accomplishing the deed. The tale recounts not the inability to foresee future events, but rather the failure to act appropriately on the knowledge available prior to their occurrence. It is both retrospective and prospective, looking backward to a moment in time in which an event was about to take place and forward in time from that moment to a future that was known. Neither the townspeople nor the authorities had stepped in to prevent the killing; the outcome was *una muerte anunciada*, a 'death foretold'.

Clearly referencing García Márquez's prescient story, two catastrophic events that converged during one week in November 1985 were both seen, in their aftermath, as *tragedias anunciadas* (or 'tragedies foretold'). On 6 November, thirty-five members of the M-19 guerrilla group attacked the Palace of Justice in central Bogotá taking hundreds of hostages, including twenty-four Supreme Court judges. President Belisario Betancur rejected their demand that he come and stand trial for accusations that he had betrayed a previously negotiated peace accord, and instead ordered the Army to storm the building. In the ensuing battle between the Colombian armed forces and the rebel gunmen, more than seventy-five hostages were killed, including eleven of the federal justices trapped inside. Exactly one week later, a volcano 80 miles west of Bogotá, the Nevado del Ruiz, erupted suddenly and nearly 30,000 people died as a massive mudslide buried the nearby town of Armero. The media quickly assigned blame for the casualties in both cases to the government's lack of foresight. Journalists demonstrated evidence showing that both events could have been anticipated and, therefore, prevented.[34] This sparked a crisis of political authority for which future-orientated security mechanisms were seen as the solution. From that moment on, the protection of human life against potential threats (both human and nonhuman in origin) would be an orientating *telos* of government: that is, a political rationality shaping the state's authority over and responsibility to its subjects. The prognostic act of forecasting the outcome of similar situations in the future – according to scientific data on the probability of natural disasters or military intelligence on the likelihood of insurgent attacks – became a key dimension of political legitimacy.

The catastrophic events of 1985 remain historical referents for contemporary understandings of political, legal, and ethical responsibility for the protection of human life in Colombia. My interviews in 2008 and 2009 with state officials and policy experts on the subject of disaster risk management were filled with references to the siege of the Palace of Justice and the Armero disaster. A social scientist involved in the creation of risk-management policy in Colombia referred to them as two of the most unforgettable events in recent memory. Like many of my other interlocutors, he pointed to them as central to the formation of political rationalities concerned with potential threats to collective life. The continued significance of these tragedies reflects how they were made to constitute a crisis in which the government's inability to anticipate and avert them was seen as a monumental failure, and predictive, preventive, and pre-emptory techniques were the solutions proposed.[35] Although Colombians were more than familiar with both political violence and natural disaster prior to November 1985, it was at this moment that a new framework emerged through which such events would be understood and managed thereafter.

This was not immediately clear to me when I began fieldwork in Bogotá on the politics of disaster risk. Although I sought to historicize what I saw as a constitutive relationship between foresight, authority, and responsibility, legitimacy, and responsibility, at first it was difficult to engage my informants in this pursuit. There was an obviousness surrounding the topic that lent it an aura of inevitability: it seemed natural to expect that the state should be able to anticipate future threats and prevent them from materializing. When I asked people to consider what may have motivated such an expectation, they often returned to 1985. For my informants, this was the year of the 'tragedies foretold'. These catastrophic events thereafter came to define the state's responsibility to protect human life from future harm.

Conclusion

I would like to conclude with a reflection on the prognostic modality of environmental politics highlighted by the Colombian and Italian cases by distinguishing between two somewhat different temporal orientations. In the first, an individual or group denounces the state for not being aware of, attentive to, or prepared for an event that has yet to take place. Back in September 2011, Doña Lucia and her neighbours looked to the future, identified a threat, organized prospective victims, demonstrated their vulnerability, and demanded pre-emptive action. The second temporal orientation, which I am calling *risk in retrospect*, involves a critique aimed at political authorities and technical experts for not *having been* aware of, attentive to, or prepared for an event that *eventually did* occur. This critique is prospective in its reference to the state's responsibility for potential threats. Yet it is also retrospective: protesters responding to the Bogotá floods and lawyers representing the L'Aquila earthquake victims both positioned themselves simultaneously in the aftermath of these events when their consequences were already known and prior to them when prevention could (or should) have been possible.[36] They mobilized a combination of hindsight and foresight to demand accountability for the loss of life and livelihood.

While the latter formulation differs from the previous one in its attempt to hold the state accountable for an actual event in the past, they share one key element: without calling into question the imperative to render the future governable in the present, those engaging in this prognostic modality of environmental politics demanded more and better mechanisms of prediction, prevention, and preparedness. And rather than disputing the fact that their status as political subjects was predicated on their vulnerability or victimhood, they sought to mobilize concern for an additional threat or to recognize a larger group of victims. Thus, while positioning themselves as critical of the state, they ultimately reinforced the future-orientated rationality of security – organized around the imperative to protect the life of the population against potential threats – as the basis of political authority and legitimacy. This raises questions about the transformative potential of prognosis in places, like Colombia, where security saturates the domain of politics and government.

These observations offer insight into the terrain of political engagement organized around environmental futures. They show that prognosis, while central to political authority, legitimacy, and responsibility, is also the ground upon which people formulate both demands on and critiques of the state. Although these claims, aspirations, and expectations may be constrained by the overarching logic of security within which they are situated, prognosis can be mobilized for a variety of political purposes. It is often deployed in order to draw attention to the state's shortcomings, contradictions, and inadequacies by demonstrating the failure to provide protective care to its most vulnerable subjects in the face of potentially catastrophic events. Across the political spectrum, the key question is how prognosis is harnessed to specific political projects and to what effect.

This is an important corrective to work that reduces the future to a domain of hegemony and domination and assumes a limited set of positions and possibilities. Temporality is not only a mechanism of social control through which to consolidate political authority, facilitate capital accumulation, and produce subjects amenable to both. In the cases I have described, prognosis is the ground upon which the state and critiques of it both rest. It is often *within* a prognostic domain, and not outside of or in opposition to it, that the poor demand more from their governments. And yet the transformative potential of prognosis as a domain of environmental politics is limited. Rather than calling into question, destabilizing, or taking hold of the state, the modality of prognosis I have analysed in Bogotá reinforces security as the orientating *telos* of government. Ultimately, while prospective denunciations take an oppositional stance *vis-à-vis* the state, they uphold the established rationalities from which they derive their meaning and force as forms of popular political expression.

I have brought together the L'Aquila earthquake and the Bogotá floods to demonstrate how prognosis becomes the basis of political engagement. But rather than revealing the 'hidden' politics of predictive techniques that are ostensibly objective, neutral, and scientific, my goal has been to highlight a modality of environmental politics in which future projections are explicitly central. And instead of showing how efforts to know and govern environmental futures shape the present, I have focused analytically on the problem of prognosis in the past. After all, concentrating exclusively on the future would risk mirroring the state's fixation on the virtual, the potential, and the possible. In contrast, focusing methodologically on the past allows us to contextualize the tight connection between foresight, authority, and responsibility. As political legitimacy is increasingly tied to knowledge about the future – of resources, of disasters, and of many other objects and events – our ability to understand the politics of prognosis becomes ever more urgent. We need to reflect critically on the strategic potential of prognosis for those usually resigned to simply wait for something better (or worse!) to come along.

NOTES

My thanks go to Jessica Barnes for her tireless efforts in assembling this collection of essays, and to her and Andrew Matthews for editorial guidance. Two anonymous reviewers also commented generously and provided helpful suggestions. An earlier version of this paper was presented at the 2012 annual meeting of the American Anthropological Association. In Bogotá, my gratitude goes to Laura Astrid Ramírez for her research assistance. The names of those quoted in the text were changed to protect their anonymity and all translations from the Spanish are my own.

[1] *http://wsp.presidencia.gov.co/Videos/2010/Noviembre/Paginas/Index.aspx* (accessed 18 January 2016). Sistema Informativo del Gobierno, Presidencia de la República de Colombia. Recording of Santos's addresses to Armero, November 2010.

[2] Prognoses of resources and disasters are additionally related by the fact that both need to account for the eventual effects of climate change. Earthquakes are different in this regard, since they are not connected to climate in any direct way. Moreover, there are really no models or simulations for predicting the occurrence of an earthquake, although such techniques can be used to predict their effects.

[3] Secondary sources consulted were the Bogotá daily newspaper *El Tiempo*, the weekly news magazine *Semana*, and the news broadcaster Noticias RCN from December 2010 to February 2012.

[4] For a related analysis of expectations surrounding innovations in the fields of heath and life science, see Brown & Michael (2003). The authors propose the concepts of 'retrospecting prospects' and 'prospecting retrospects' as 'interpretive registers' through which people understand and discuss expectations of future change.

[5] For similar diagnoses of temporal politics in urban studies, see Auyero & Swistun (2009), Mitchell (2009), and Yiftachel (2009).

[6] An inspiration for this analysis is Koselleck, who writes 'not of one historical time, but rather of many forms of time superimposed on one another' (2004: 2).

[7] Here I am in agreement with Adams *et al.* when they conclude their critique with the question: 'What would it mean to not-anticipate?' (2009: 260). Although their article seeks to imagine 'strategies of refusal' that disrupt anticipatory regimes, they nevertheless recognize that 'perhaps a better tactic is not to refuse anticipation' and instead consider what relations to the future are desirable, and how we might go about fostering them.

[8] Gisa Weszkalnys's (2014) work offers another important resource for thinking about anticipation and the temporal politics of disaster. See also her contribution to this volume.

[9] These anticipatory practices hinge on the recognition that dominant logics of capital accumulation and political rule are always fractured and inconclusive – as Simone argues, they are 'full of potential holes capable of providing, albeit always temporarily, shelter and maneuverability' (2010: 98).

[10] After all, it is also through the promise of alternative futures that members of the urban poor come to believe that political change or economic development will eventually materialize, which limits their options and reduces their leverage.

[11] For discussion of the relationship between past, present, and future, see Huyssen (2003), Koselleck (2004), and Luhmann (1998).

[12] We are now aware that this idea was predicated on colonial assumptions about time and space, such that futurity mapped on to the spatial categories of centre/periphery and West/non-West. Geographical distance was equated with temporal difference (Fabian 2002), as places beyond the metropole were relegated to a time before the present.

[13] Naomi Klein (2007) traces the political-economic logic of what she calls 'disaster capitalism'.

[14] I am grateful to an anonymous reviewer for drawing connections between my argument about the state's responsibility to predict and mitigate future catastrophes and James Laidlaw's (2013: 204-9) work on the proliferation of ethical responsibilities enabled by new forms of statistical reasoning.

[15] *Semana*, 7 December 2011; *El Tiempo*, 8 December 2011.

[16] *Semana*, 9 December 2011.

[17] 'Autorizan inundaciones controladas en baldíos del norte y sur de Bogotá', *Semana*, 7 December 2011.

[18] 'Damnificados subieron a 45.196 en tres días', *El Tiempo*, 9 December 2011.

[19] 'Inundaciones: gobierno entregará ayuda económica a damnificados en Bogotá', *Semana*, 9 December 2011.

[20] Ibid.; 'Ayuda para damnificados de Bosa será entregada la próxima semana', *El Tiempo*, 10 December 2011.

[21] 'En Bosa hay 5 mil damnificados a causa de las fuertes lluvias', *El Tiempo*, 8 December 2011; 'Damnificados subieron a 45.196 en tres días', *El Tiempo*, 9 December 2011.

[22] 'Bogotá: así se entregan las ayudas en Bosa y Kennedy', *Semana*, 13 December 2011.

[23] 'Víctimas de invierno en Bosa y Kennedy viven drama para recibir ayudas', *El Tiempo*, 13 December 2011.

[24] 'Damnificados subieron a 45.196 en tres días', *El Tiempo*, 9 December 2011.

[25] 'Después del diluvio', *Semana*, 7 January 2012.

[26] *El Tiempo*, 7 December 2011; *Semana*, 7 December 2011.

[27] 'Después del diluvio', *Semana*, 7 January 2012.

[28] Yolanda Reyes, 'No es la "maldita niña"', *El Tiempo*, 12 December 2011.

[29] 'La alcaldesa respondió a las denuncias por la entrega de ayudas', *El Tiempo*, 14 December 2011.

[30] Ernesto Cortés Fierro, 'Alcalde, ¿y del invierno . . . ?', *El Tiempo*, 11 December 2011.

[31] 'Al invierno, se suma drama para recibir ayudas', *El Tiempo*, 13 December 2011.

Journal of the Royal Anthropological Institute (N.S.), 163-180
© Royal Anthropological Institute 2016

[32] 'Antonio Navarro Wolff asumió como Secretario de Gobierno de Bogotá', *El Tiempo*, 2 January 2012; 'Con modelo hidráulico buscan evitar nuevas inundaciones en Bogotá', *Semana*, 17 January 2012.

[33] Noticias RCN, 9 December 2011, *http://www.canalrcnmsn.com/noticias/inundación_en_bosa_hab%C3% AD_sido_denunciada_desde_hace_tres_meses* (last accessed 23 May 2014).

[34] *El Tiempo*, 16 November 1985; Daniel Samper Pizano, 'Apocalipsis anunciado', *El Tiempo*, 18 November 1985; Gloria Moanack, 'Colombia, region de alto riesgo sísmico', *El Tiempo*, 14 November 1985.

[35] For a discussion of the *crisis de 'gubernabilidad'* that ensued after the coincidence of the Armero tragedy and the attack on the Palace of Justice, see Ramírez Gomez & Cardona (1996: 267).

[36] To quote one of the lawyers for the victims' families: 'It's not possible to predict an earthquake. But it was possible to predict the seismic risk in L'Aquila after months of tremors'. As such, this modality of prognostic politics often supports demands for emergency response, humanitarian aid, and financial reparations.

REFERENCES

ADAMS, V., M. MURPHY & A.E. CLARKE 2009. Anticipation: technoscience, life, affect, temporality. *Subjectivity* **28**, 246-65.

AUYERO, J. & D.A. SWISTUN 2009. *Flammable: environmental suffering in an Argentine shantytown*. New York: Oxford University Press.

BAKHTIN, M. 1981. *The dialogic imagination: four essays* (trans. M. Holquist). Austin: University of Texas Press.

BEAR, L. 2014. Doubt, conflict, mediation: the anthropology of modern time. *Journal of the Royal Anthropological Institute* (N.S.) **20**, 3-30.

BROWN, N. & M. MICHAEL 2003. A sociology of expectations: retrospecting prospects and prospecting retrospects. *Technology Analysis & Strategic Management* **15**, 3-18.

BUCK-MORSS, S. 2000. *Dreamworld and catastrophe: the passing of mass utopia in East and West*. Cambridge, Mass.: MIT Press.

COOPER, M. 2006. Pre-empting emergence: the biological turn in the War on Terror. *Theory, Culture & Society* **23**, 113-35.

——— 2010. Turbulent worlds: financial markets and environmental crisis. *Theory, Culture & Society* **27**, 167-90.

EDWARDS, P.N. 2010. *A vast machine: computer models, climate data, and the politics of global warming*. Cambridge, Mass.: MIT Press.

——— 2012. Scaling disaster: simulating the extent of the Deepwater Horizon and the Fukushima meltdowns. Society for Social Studies of Science annual meeting, Copenhagen, 20 October 2012.

ESCOBAR, A. 1992. Imagining a post-development era? Critical thought, development and social movements. *Social Text* **31/32**, 20-56.

FABIAN, J. 2002. *Time and the other: how anthropology makes its object*. New York: Columbia University Press.

FERGUSON, J. 2006. *Global shadows: Africa in the neoliberal world order*. Durham, N.C.: Duke University Press.

FUKUYAMA, F. 1992. *The end of history and the last man*. New York: Free Press.

GARCÍA MÁRQUEZ, G. 1981. *Crónica de una muerte anunciada*. Bogotá: La Oveja Negra.

GEERTZ, C. 1973. *The interpretation of cultures*. New York: Basic Books.

GELL, A. 1996. *The anthropology of time: cultural constructions of temporal maps and images*. Oxford: Berg.

GREENHOUSE, C.J. 1996. *A moment's notice: time politics across cultures*. Ithaca, N.Y.: Cornell University Press.

HUYSSEN, A. 2003. *Present pasts: urban palimpsests and the politics of memory*. Stanford: University Press.

KLEIN, N. 2007. *The shock doctrine: the rise of disaster capitalism*. New York: Metropolitan Books/Henry Holt.

KOSELLECK, R. 2002. *The practice of conceptual history: timing history, spacing concepts* (trans. T.S. Presner *et al.*). Stanford: University Press.

——— 2004. *Futures past: on the semantics of historical time* (trans. K. Tribe). New York: Columbia University Press.

LAIDLAW, J. 2013. *The subject of virtue: an anthropology of ethics and freedom*. Cambridge: University Press.

LEVITAS, R. 1982. Dystopian times? The impact of the death of progress on utopian thinking. *Theory, Culture & Society* **1**, 53-64.

LUHMANN, N. 1998. *Observations on modernity*. Stanford: University Press.

MARTIN, R. 2007. *An empire of indifference: American war and the financial logic of risk management*. Durham, N.C.: Duke University Press.

MASSUMI, B. 2009. National enterprise emergency: steps toward an ecology of powers. *Theory, Culture & Society* **26**, 153-85.

——— 2010. Perception attack: brief on war time. *Theory & Event* **13**.

MITCHELL, K. 2009. Pre-black futures. *Antipode* **41**, 239-61.

MUNN, N.D. 1992. The cultural anthropology of time: a critical essay. *Annual Review of Anthropology* **21**, 93-123.

PINDER, D. 2005. *Visions of the city: utopianism, power, and politics in twentieth-century urbanism.* New York: Routledge.

RABINOW, P. 2008. *Marking time: on the anthropology of the contemporary.* Princeton: University Press.

RAMÍREZ GOMEZ, F. & O.D. CARDONA 1996. El sistema nacional para la prevención y atención de desastres en Colombia. In *Estado, sociedad y gestión de los desastres en América Latina: en busca del paradigma perdido* (eds) A. Lavell & E. Franco, 255-307. Lima: La RED, FLACSO, ITDG-Perú.

ROBINSON, J. 2004. In the tracks of comparative urbanism: difference, urban modernity and the primitive. *Urban Geography* **25**, 709-23.

ROSENBERG, D. & S. HARDING 2005. Introduction: histories of the future. In *Histories of the future* (eds) D. Rosenberg & S. Harding, 3-18. Durham, N.C.: Duke University Press.

ROY, A. 2008. The 21st-century metropolis: new geographies of theory. *Regional Studies* **43**, 819-30.

SIMONE, A. 2010. *City life from Jakarta to Dakar: movements at the crossroads.* London: Routledge.

WESZKALNYS, G. 2014. Anticipating oil: the temporal politics of a disaster yet to come. *The Sociological Review* **62**, 211-35.

YIFTACHEL, O. 2009. Theoretical notes on 'gray cities': the coming of urban apartheid? *Planning Theory* **8**, 88-100.

ZEIDERMAN, A. 2012. On shaky ground: the making of risk in Bogotá. *Environment and Planning A* **44**, 1570-88.

——— 2013. Living dangerously: biopolitics and urban citizenship in Bogotá, Colombia. *American Ethnologist* **40**, 71-87.

ŽIŽEK, S. 2010. *Living in the end times.* New York: Verso.

Pronostiquer le passé : politique temporelle des catastrophes en Colombie

Résumé

L'auteur explore, dans le présent article, une modalité pronostique de la politique environnementale qu'il appelle *risque rétrospectif*. Pour cela, il examine des conflits liés non seulement à la capacité de la science d'État à connaître et régir le futur, mais aussi à son incapacité d'y parvenir par le passé. Il commence par discuter du séisme qui a frappé la ville italienne de L'Aquila en 2009, avant de se tourner vers un cas similaire en Colombie, reflétant l'un comme l'autre la relation de constitution entre autorité politique et prévision. Il remonte ensuite au fort précédent historique de ce type de situation politique en Colombie en discutant d'une référence culturelle essentielle : le livre de Gabriel García Márquez *Chronique d'une mort annoncée*, paru en 1981, ainsi que de la convergence de deux événements catastrophiques survenus la même semaine en novembre 1985 et dont les conséquences ont été qualifiées de « tragédies annoncées ». Pour finir, il réfléchit aux implications possibles de l'éventualité où le pronostic deviendrait le terrain sur lequel les citoyens s'engagent dans des relations politiques avec l'État.

10

Claiming futures

Elizabeth Ferry *Brandeis University*

The papers in this special issue on 'environmental futures' draw liberally on cross-disciplinary conversations, yet their strength comes from their ethnographic depth and their characteristically anthropological willingness to consider diverse types of entities and phenomena in the same holistic frame. The range of places (Oman, Alaska, Egypt, Colombia, Antarctica, etc.) and entities (ice, oil, gold, governmental officials, glaciologists, salmon, models and scenarios, PowerPoint presentations, rainfall, etc.) engaged in these instances of 'prognostic politics' provide the kind of material that distinguishes the anthropological project of building theory through ethnography and comparison. In writing this response, I group the papers in pairs under four topics that bring out some of the especially interesting and novel contributions of the issue as a whole. The themes are: temporality and uncertainty; anticipatory knowledges; resource affect; and material signs. These themes refract the visions presented in the papers, showing details of the process of claiming multiple uncertain, agonistically engaged futures, and the consequences of these claimed futures. I briefly conclude by considering these papers as part of the current pragmatist (sometimes called 'ontological') bent of some anthropological work, and the heuristic possibilities provided by this orientation.

The papers in this special issue on 'environmental futures' focus on resources and disasters as temporally and materially embedded phenomena, and on the temporal and material political fields they generate. The essays attend to these phenomena as constitutive of a contemporary form of politics described by Andrew Mathews and Jessica Barnes in their introduction as 'prognostic politics'. This emergent (though of course not unprecedented) mode of politics is distinguished by participants' recognition of multiple visions of the future, the hybridity of the forums in which politics is played out, and the explicitness with which the boundaries of nature and the social are traversed.

The essays draw liberally on cross-disciplinary conversations, such as new literature in political ecology informed by science and technology studies (Goldman, Nadasdy & Turner 2010; Kosek 2010), social studies of risk and uncertainty (Beck 1992; Luhmann 1993; Zaloom 2004), and ontological approaches to the analysis of material/social processes (Blaser 2013; Holbraad 2012; Mol 2002; Pedersen 2011). Yet they approach these questions through the methods and outlook of anthropology; much of the pieces' strength comes from their ethnographic depth and their willingness to consider diverse types of entities and phenomena in the same holistic frame. The range of places (Oman, Alaska, Egypt, Colombia, Antarctica, etc.) and entities (ice, oil, gold, governmental officials, glaciologists, salmon, models and scenarios, PowerPoint presentations, rainfall, etc.) engaged in these instances of prognostic politics provide the kind of material that distinguishes the anthropological project of building theory through ethnography and comparison.

In this commentary, I do not attempt to sum up the essays in any global or comprehensive way. Rather, I present my own partial and positioned response, drawing out some of themes and resonances that I found most exciting and intriguing. I do so by grouping the papers in pairs under four topics (clearly not the only possible ones), indicating connections across topics along the way. These themes bring out some of the especially interesting and novel contributions of the issue as a whole. The themes are: temporality and uncertainty; anticipatory knowledges; resource affect; and material signs.

I choose the title phrase 'Claiming futures' to draw attention to two prominent features of the essays: claims as building blocks in scientific arguments central to prognostic politics; and claims as parcels of land or water over which a claimant has the legal right to extract resources. These different sorts of claims correspond to the different versions of the present and future, materialized in land, samples of ice or rock, computer models, and documents, that circulate within politics surrounding environmental futures.

Temporality and uncertainty: disasters foretold

Austin Zeiderman's paper on the politics surrounding flooding in December 2011 in the Bosa area on the outskirts of Bogotá, Colombia, explores what he describes as 'a peculiar temporality – at once prospective and retrospective – that often characterizes public controversies in the aftermath of catastrophe' (p. 156). Because residents of the neighbourhood had pointed out the risk of flooding a few months earlier, the local and federal authorities were seen by many as culpable, not only of responding incompletely to the disaster, but of failing to take precautions against it. As Zeiderman suggests, this form of reasoning depends on a particular set of relations between present and future, perhaps like a kind of multiple predestination. It is as though there are multiple possible futures tied to any present moment. A given negative prediction is a sign not that a given future is inevitable, but that it is sufficiently possible that preparations should be made for it. Moreover, from the perspective of that future post-catastrophic moment, the link is now inevitable. At that later moment, it appears as though nothing else could have happened, because of the one thing that did in fact happen. One among the possible futures that were imaginable has been actualized, carrying culpability with it.

In an article theorizing different forms and technologies of uncertainty based on ethnographic work on pandemic preparedness in Israel, Limor Samimian-Darash describes this moment of multiple possible futures as 'potential uncertainty', which 'is not equivalent to the unknown future but is linked to the intermediate space between what has occurred and what is about to occur. In this space, a potential exists for the appearance of various actualities. Potential uncertainty denotes the opening for a variety of possibilities' (2013: 3). Where Samimian-Darash delineates the technologies developed by the Israeli government to cope with the situation of potential uncertainty, Zeiderman documents what happens once that moment of potential has been closed down by the actualization of one possibility over others (Goldstein 2013). He shows, further, how the relationship between the earlier space of potential uncertainty and the later one of actualization becomes the ground for a retrospectively prognostic politics.

Karen Hébert's exploration of the participatory public discussions surrounding the Pebble Mine and its potential effects in the Bristol Bay region of Alaska also documents the political spaces that are opened up and closed down in a vision of multiple possible worlds that can only be imperfectly predicted. In Hébert's essay, these possible futures are enacted by scenarios for the possible effects of the proposed gold, copper, and molybdenum mine, known as the Pebble Mine, that developers and investors hope to construct in Bristol Bay. In the context of unusual vulnerability of the mining corporation because of the particular status of Bristol Bay as one of the last intact breeding grounds for salmon and also because of falling metals prices, risk scenarios developed by the mine's opponents and by the US Environmental Protection Agency (EPA) are, in Hébert's words, 'saturated with risk' (p. 113). Thus Hébert's paper describes a similar temporal frame to Zeiderman's but from an earlier vantage-point, one that poses Bristol Bay as '[s]o pristine as to seem almost prelapsarian, the area emerges from this imagery as a not-yet-ruined world (p. 110). The fact that the disaster is still being foretold creates a different field for prognostic politics, one in which science and politics are continually intertwined in what Michel Callon and his colleagues have described as 'hybrid forums' (Callon, Lascoumes & Barthe 2011; Callon, Méadel & Rabeharisoa 2002).

Hébert's argument focuses especially on the ways in which scientific and other forms of expertise are deployed in the spaces opened up within this particular domain of risk assessment. She notes that scientific expertise is both mobilized and called into question in different venues for debate over the mine's predicted impact. The temporal positioning of the present as jumping-off point for multiple possible futures provides the counterpoint for Zeiderman's depiction of retrospective politics in post-flood Bogotá, where government officials are called to account for neglecting the future that actually came to pass. Both essays thus invoke a particular relationship between time and politics, by positing a set of multiple, uncertain, politically saturated futures, of which only one has (or will have) come to pass.

Anticipatory knowledges

The papers by Jessica Barnes and Jessica O'Reilly also engage the question of multiple futures, this time through climate change models of different sorts. O'Reilly looks at several types of glaciologists, including those gathering data in the field (on the ice) and those constructing models at a distance. She explores these scientists' engagements with Antarctic ice, particularly the Western Antarctic Ice Sheet (WAIS), which has been a matter of scientific and public concern for some decades. She addresses the question of uncertainty and the projection of possible future worlds through diverse forms of knowledge and expertise. Glaciologists who interact with ice directly have particular ways of knowing the ice; at the same time the data they collect contribute to more geographically distant but in some ways equally intimate knowledges, in the form of inputs for future models of ice behaviour over the coming decades and centuries.

Since there is little consensus on what exactly will happen to WAIS and when, scientists find themselves in various quandaries over what versions of the future to create and ratify. By looking at modelling of ice and interacting with ice in other ways (such as walking on it, observing its behaviour, experiencing or avoiding its dangers),

O'Reilly broadens the terrain on which present knowledge casts up possible futures, by showing how, in her words, 'scientists working in future time fill in gaps in their predictive capabilities with expertise of the ice sheet that is both intimate and technical' (p. 32).

Jessica Barnes' essay on how Egyptian and international scientists model the future of the Nile's water flow also explores the productive nature of uncertainty. She notes that while many scientists believe that the representations of precipitation in East Africa made by general circulation models (GCMs) are not sufficient for confident predictions of what exactly future water flows will be in the Nile, the models continue to produce forms of knowledge and to provide the material for prognostic politics. As in O'Reilly's case, what uncertainty makes more and less possible varies for different scientists, in this case along national lines. Egyptian scientists are more likely to accept uncertain scenarios as useable than are other scientists, and more likely to accept those that predict less rainfall in the future than those that predict more. This phenomenon resonates strongly with the experiences described in Hébert and O'Reilly's papers as well, where politics and science are interwoven at every level. Within these hybrid forums, models of the future do not necessarily presuppose (at least not for the scientists involved) that increased knowledge will drive out uncertainty, though there is still some deference given to scientific expertise. We can see this in the public discussion over predicted effects of the Pebble Mine on salmon, described by Hébert, and perhaps also underlying the laughter that Barnes has observed in non-scientific audiences when they hear that scientific models predict that the Nile's flow will either increase or decrease in the future (since the laughter seems to come from the assumption among non-scientists that science is for increasing certainty about the future).

Resource affect: seeping through time

Gisa Weszkalnys and Mandana Limbert both treat the production of possible futures in the context of oil extraction, and perhaps because of this particular material resource, their papers concern a different register of expectations, hopes, dreams, and dreads. Weszkalnys uses the term 'resource affect' to describe the intensity that seeps through experiences surrounding resources, often spilling over the boundaries of traditional political economy accounts. Weszkalnys notes that '[i]t is … the ethnographer's task to chart the specificity of this attachment [of affect] in its different cultural and historical inflections' (p. 119). She approaches this task through attention to oil's double promise/threat of generativity and destruction in the context of São Tomé and Príncipe (STP). Weszkalnys captures not only the affect of resource histories and futures, but also the historicity and futurity of affect.

Weszkalnys describes the prospective oil economy in STP as something that is generating considerable hopes and fears, in the absence of any oil having actually been extracted. STP has existed in a drawn-out, seemingly unending period of exploration ever since possible offshore oil reserves were announced in the late 1990s. The persistent presence of future oil and present no-oil has itself generated multiple affective orientations towards the future, including hope, resignation, and 'doubtful hope', as Weszkalnys's informant Lourdes puts it.

In the case of Oman's oil production, examined in Mandana Limbert's paper, we find a contrasting situation: oil is being extracted but is absent in the discursive enactment of the Omani nation. Limbert shows how the importance of oil is downplayed in textbooks and other national histories, and is at times substituted by a history centred on the problem of succession (since the present Sultan has no heir) and the possible return of the Imamate. To the occlusion of oil's appearance in historical accounts and the downplaying of its agency, Limbert counterposes the ways that water's appearance and disappearance is 'not only described as a harbinger of divine pleasure and anger, but is also a miraculous motivator and actor' (p. 140). The contrast between these two so-called '"natural" liquids that help produce wealth and prosperity' (p. 141) reveals further nuance in the affective materialities of resources.

Furthermore, where in STP the beginning of production is continually postponed, in Oman it is the close of the oil age when reserves are depleted that is pushed into the future (reserves have been projected to last for twenty years for the past four decades). However, the two cases are alike in one way, particularly significant to the aims of the special issue: in both Oman and STP, resource affects produce ambivalent futures, whether seen in terms of oil's emergence or its depletion. Through an attention to those affective, substantial aspects of oil and water that tend to permeate political ecology and economy discussions, Weszkalnys and Limbert show how the affective dimensions of resource-making projects (Ferry & Limbert 2008) also saturate people's conceptions of time and the future. Indeed, while affect is certainly present in other essays in this special issue, perhaps most notably in Jessica O'Reilly's discussion of glaciologists' perception of melting ice, it is perhaps significant that resource affect, with its seeping, soaking, brimming qualities, seems especially evident in these two papers on oil.

Material signs

The papers by David Kneas and Nusrat Chowdhury both deal with mined material (drill cores and lumps of coal) as material signs that are intended to convey meaning to particular audiences (such as investors, citizens, or activists). In each case, the claim of unified semiotic legibility disintegrates into several possible readings within a contentious political field. David Kneas examines 'prognosis' in two aspects of the work of Ascendant Copper, a now defunct junior mining company that aimed to build a mine in the Intag region in Ecuador. Ascendant's positioning with respect to its investors and to the surrounding community exemplified several aspects of the 2000s minerals book: the need to represent reserves in the era after the Bre-X gold hoax; and the corporate social responsibility movement, in which mining companies were compelled to represent themselves as responsible development actors, in this case in contrast to the supposedly destructive forms of slash-and-burn agriculture in the Intag zone.

The drill cores collected by Bishimetals, Ascendant's predecessor, appear at first to have a straightforward indexical relationship to the copper lying underground, which in the post-Bre-X system is classified according to scale of measured, indicated, and inferred resources. This scale attempts to account for uncertainty, but also, as Kneas points out, opens possibilities for companies to generate a whole range of claims that can then be strategically used to generate interest for investors and possible buyers. Thus, the indexicality by which the drill cores are intended to stand for the deposit as a whole does not resolve but rather produces uncertainties, which are themselves productive, within what Kneas describes as 'a temporal politics of "in the meantime" – a time period during which Ascendant could explore the Junin deposit without being questioned about what that exploration would lead to' (p. 68).

In Chowdhury's essay, we also find the semiotic production of uncertainty, in this case through the iconic relationship between the shininess of the coal that her informant

Ranjan observes emerging from the ground, about which he says, 'We have been seeing very high-quality coal. It shines' (p. 89). The shininess is the qualisign of value for this observer (Munn 1992; Peirce 1955 [1902]): that is, it signifies value because value is also brilliant, as with 'gold, precious gems, and other objects with surfaces that reflect light' (p. 89).

However, iconicity in Chowdhury's essay is no more reliable as a semiotic process than indexicality in Kneas's, if by reliability we mean the production of a unified and stable meaning. For as Ranjan points out, using the double meaning of the verb 'to see' as a pivot, there may be other unseen values behind the shininess of coal. The fact that the 'ordinary folks' are kept back from the site by guards and red ribbons suggests to Ranjan that the coal's shininess may hide value as well as reveal it. The uncertainty produced in this scene of emergence and possible withholding provides the ground for activism surrounding the mine, what comes out of it, and where its profits will go. As Chowdhury argues, 'The unintended slippages in meanings, the inadvertent confusion of words, and the consequent hazards of representation … can be, and often are, significant sources – or resources – of political action' (p. 93).

The counterpoint of indexicality and iconicity as modes of semiosis in Chowdhury's and Kneas's papers suggests an avenue for further discussion of prognosis politics. Readers and readings of signs proliferate in these essays, including glaciologists, residents of flood-prone areas, GCMs, drill cores, EPA reports, and Alaskan hunters. Claims are made on the basis of contiguity and context (indexicality) and similarity and analogy (iconicity), and the legitimacy of these claims often depends on assumptions of how these semiotic connections work. A deeper analysis into these semiotic infrastructures would tell us a great deal about how different environmental futures are generated, circulated, and received within the hybrid forums described in these essays.

Concluding thoughts

Through these contributions, we can see the details of the process of claiming multiple uncertain, agonistically engaged futures, and the consequences of these claimed futures, in an ongoing process of what Mario Blaser (2013) has described as 'ontological politics' or 'worlding' (also see de la Cadena 2010; Mol 2002, among many other examples). In keeping with the pragmatist bent of much recent anthropological discussion, the essays in this issue tend to bracket the question of how accurately climate models or other material semiotic objects involved in prognostic politics do in fact predict the future, and focus instead on how these models or objects work and what they do. In doing so, they tap into broader discussions within anthropology, worth noting here. On the one hand, Paul Edwards (2010), Kirsten Hastrup (2013), and others have closely examined the protagonism of climate modelling *per se*, and several essays in this issue draw on and extend that work. The essays in this issue also draw on broader discussions concerning the pragmatics of documents and other objects that circulate through political fields to generate new assemblages and form new paths of translation (Latour 2007 [among other works]; Mathews 2011; Riles 2001). These approaches privilege the ways that objects act and interact over the ways that they represent, as the participants in the political fields in which these objects circulate often do themselves (Ferry & Wood 2012; Hetherington 2011; Miyazaki 2013). Though some argue that this move is merely a dressed-up version of what the concept of culture has been doing all along (Carrithers 2010; Keane 2013), I would argue that it provides heuristic leverage that has opened up deadlocked conversations about science, nature, and politics (Latour 2004). The essays in this issue demonstrate that this heuristic move allows for both the enactment and the analysis of prognostic politics.

In taking this approach, the essays build on recent conversations spanning disciplines. However, in their detailed ethnography and in their examination of the ways that prognostic politics are saturated with time, materiality, and affect, the essays in this special issue enrich these cross-disciplinary discussions on the ontological politics of the future with descriptions that are textured, located, and at once particular and comparative. They thus bring the classic strengths of anthropology to the table, allowing for a more layered, multi-perspectival, and located understanding of the politics surrounding environmental futures.

REFERENCES

BECK, U. 1992. *Risk society: towards a new modernity* (trans. M. Ritter). London: Sage.

BLASER, M. 2013. Ontological conflicts and the stories of peoples in spite of Europe: towards a conversation on political ontology. *Current Anthropology* **54**, 547-68.

CALLON, M., P. LASCOUMES & Y. BARTHE 2011. *Acting in an uncertain world: an essay on technical democracy.* Cambridge, Mass.: MIT Press.

CALLON, M., C. MÉADEL & V. RABEHARISOA 2002. The economy of qualities. *Economy and Society* **31**, 194-217.

CARRITHERS, M. 2010. Ontology is just another word for culture. *Critique of Anthropology* **30**, 152-200.

DE LA CADENA, M. 2010. Indigenous cosmopolitics in the Andes: conceptual reflections beyond 'politics'. *Cultural Anthropology* **25**, 334-70.

EDWARDS, P.N. 2010. *A vast machine: computer models, climate data, and the politics of global warming.* Cambridge, Mass.: MIT Press.

FERRY, E.E. & M.E. LIMBERT 2008. Introduction. In *Timely assets: the politics of resources and their temporalities* (eds) E.E. Ferry & M.E. Limbert, 3-24. Santa Fe, N.M.: School for Advanced Research Press.

——— & D.C. WOOD 2012. Expert vocabularies: metrics and indicators as forms of pragmatic truth. Paper presented at Initiative on Climate Adaptation Research and Understanding through the Social Sciences (ICARUS) 3rd Annual Conference, 18 May.

GOLDMAN, M.J., P. NADASDY & M.D. TURNER (eds) 2010. *Knowing nature: conversations at the intersection of political ecology and science studies.* Chicago: University Press.

GOLDSTEIN, D. 2013. Comment on L. Samimian-Darash, 2013. Governing future potential biothreats: toward an anthropology of uncertainty. *Current Anthropology* **54**, 14-15.

HASTRUP, K. 2013. Anticipating nature: the productive uncertainty of climate change models. In *The social life of climate change models: anticipating nature* (eds) K. Hastrup & M. Skrydstrup, 1-29. New York: Routledge.

HETHERINGTON, K. 2011. *Guerrilla auditors: the politics of transparency in neoliberal Paraguay.* Durham, N.C.: Duke University Press.

HOLBRAAD, M. 2012. *Truth in motion: the recursive anthropology of Cuban divination.* Chicago: University Press.

KEANE, W. 2013 Ontologies, anthropologists, and ethical life. Comment on G. Lloyd, 2012. *Being, humanity, and understanding.* Oxford: Oxford University Press. *Hau: A Journal of Ethnographic Theory* **3**, 186-91.

KOSEK, J. 2010. Ecologies of empire: on the new uses of the honeybee. *Cultural Anthropology* **25**, 650-78.

LATOUR, B. 2004. Why has critique run out of steam? From matters of fact to matters of concern. *Critical Inquiry* **30**, 225-48.

——— 2007. *Reassembling the social: an introduction to Actor-Network-Theory.* Oxford: Clarendon Press.

LUHMANN, N. 1993. *Risk: a sociological theory.* Boston: Walter de Gruyter.

MATHEWS, A.S. 2011. *Instituting nature: authority, expertise, and power in Mexican forests.* Cambridge, Mass.: MIT Press.

MIYAZAKI, H. 2013. *Arbitraging Japan: dreams of capitalism at the end of finance.* Berkeley: University of California Press.

MOL, A. 2002. *The body multiple: ontology in medical practice.* Durham, N.C.: Duke University Press.

MUNN, N. 1992. *The fame of Gawa: a symbolic study of value transformation in a Massim society.* Durham, N.C.: Duke University Press.

PEDERSEN, M.A. 2011. *Not quite shamans: spirit worlds and political lives in northern Mongolia.* Ithaca, N.Y.: Cornell University Press.

PEIRCE, C.S. 1955 [1902]. *Three trichotomies of signs: philosophical writings of Peirce* (ed. J. Buchler). New York: Dover.

RILES, A. 2001. *The network inside out.* Ann Arbor: University of Michigan Press.

SAMIMIAN-DARASH, L. 2013. Governing future potential biothreats: toward an anthropology of uncertainty. *Current Anthropology* **54**, 1-22.

ZALOOM, C. 2004. The productive life of risk. *Cultural Anthropology* **19**, 365-91.

Revendiquer des futurs

Résumé

Les articles de ce numéro spécial sur « les futurs de l'environnement » puisent largement à la source de conversations interdisciplinaires. Leur force vient cependant de leur profondeur ethnographique et de leur disposition typiquement anthropologique à étudier différents types d'entités et de phénomènes dans un même cadre global. La diversité des lieux (Oman, Alaska, Égypte, Colombie, Antarctique, etc.) et des entités (glace, pétrole, fonctionnaires, glaciologues, saumon, modèles et scénarios, présentations PowerPoint, pluviométrie, etc.) impliqués dans ces cas de « politique des pronostics » produit un matériau qui fait la singularité du projet anthropologique de théorisation par l'ethnographie et la comparaison. Dans cette réponse, je regroupe les articles par paires sous quatre rubriques qui mettent en lumière quelques apports particulièrement intéressants et nouveaux de ce numéro dans son ensemble : temporalité et incertitude, connaissances anticipatrices, affect des ressources et signes matériels. Ces thèmes réfractent les visions présentées dans les articles, révélant les détails du processus de revendication de multiples futurs incertains et en lutte et les conséquences de ces futurs revendiqués. Je conclus rapidement en considérant ces articles comme une composante du virage pragmatiste (parfois appelé « ontologique ») de certains travaux anthropologiques et en examinant les possibilités heuristiques que révèle cette orientation.

Index

access to data, 57–8
Actor Network Theory, 148
Adams, Vincanne, 89, 96, 102, 166, 178
agency, 132, 148; corporate, 77, 82
agents, 13, 80, 91, 94, 131
Alaska, 14, 19, 108–26
Antarctica, and climate change, 40;
  and research interests, 43; and scientific
  expertise, 27–45, 183; crevasses,
  31–3
Anthropocene age, 20–1
anticipation, 10, 89, 160; anticipatory
  knowledge, 183–4; anticipatory politics,
  166–7
apocalypse, 9, 160, 167
Appadurai, Arjun, 89, 102
atomic energy, 11
Badiou, Alain, 149, 160
Bakhtin, Mikhail, 167
Bangladesh, and energy, 17, 19, 87–107;
  services sector, 96
Barad, Karen, 68, 69
Barnes, Jessica, vi, 9–26, 46–66, 181, 184
Beck, Ulrich, 10, 112, 113, 181
Benson, Peter, 69
Berlant, Lauren, 149
biodiversity, 9, 78
biogas, 79

Blaser, Mario, 186
Buck-Morss, Susan, 167

Callon, Michel, 20, 110, 112, 123, 183
Canadian Institute of Mining, 73–4
Cannetti, Elias, 133
Cash, Marc, 35–6
causality, 147, 148, 149, 159
Chander, Krishan, 100
Chowdhury, Nusrat Sabina, vi, 17, 19, 22,
  87–107, 185–6
Choy, Timothy, 113
cities, 167, 168; green, 9
climate change, 9, 11, 12, 18, 31, 40, 166,
  178; and local perceptions, 62; and the
  Nile, 46–7, 50–1, 53, 55–6; IPCC, 39–40;
  knowledge, 48; prediction, 12, 48; see
  also greenhouse gases; ice melt
climate science, 10, 11, 29; models, 13, 14,
  19, 21, 29, 41, 57, 166, 186
Clinton, Bill, 95
coal, 17, 18, 19, 87–107; quality, 98–9, 185–6
coal mining, protests, 87–8, 97, 100–01
Cold War, 11, 13, 38, 167
Colombia, and flooding, 18, 20, 21,
  168–73; and environmental politics,
  163–80; *El Espectador*, 172; *El Tiempo*,
  168, 170, 172, 178

compensation, 88, 95, 104, 128, 132, 133
conflict, 59; anti-mining, 70, 83, 87–8, 89, 92, 97
Coronil, Fernando, 123, 129
corporate social responsibility, 69–70, 77, 80, 81, 82, 83, 90, 103
Cross, Jamie, 89

Darwin, Charles, 29
deforestation, 9, 77, 78, 80; REDD, 14
Deleuze, Gilles, 160
development, 15–6, 67, 68, 78, 80, 81, 82, 89, 97, 102, 129, 167, 185; Millennium Development Goals, 16; post-development, 167; Sustainable Development Goals, 16
diagrammatic reasoning, 12
disasters, 14, 15, 18, 119, 163–80; scenarios, 113–5

earthquakes, 20, 163–4, 165, 168, 174, 176, 177, 178, 179
Ecuador, 13, 19, 67–86; subsoil copper wealth, 67, 71–84
Edwards, Paul, 13, 56, 57, 165–6, 186
Egypt, and river flow, 18, 19, 21, 46–66, 184
ethnographic research, 27, 37, 71, 83, 108, 111, 127
Evans-Pritchard, E.E., 9, 23
expert elicitation, 37–40, 42; and IPCC, 39–40; Delphi method, 38

Ferguson, James, 134, 140, 141, 148, 167
Ferry, Elizabeth, vi, 20, 21, 113, 120, 128, 181–8
fetishization, 95, 143
fisheries, 10, 108, 110, 114, 118–9
food security, 91
forests, 10, 12, 16, 78, 91
Foucault, Michel, 16–7, 77, 80, 83

García Márquez, Gabriel, *Chronicle of a death foretold*, 165, 174–5
geopolitics, Nile basin, 50–1
glaciologists, 13, 27–45, 183, 185
Gluckman, Max, 149
gold rush, 131

green cities, 9
greenhouse gases, 14, 47, 51, 56
Greenland, 12, 34, 41
greetings cards, 93, 94

Hastrup, Kirsten, 12, 186
Hébert, Karen, vi, 14, 19–20, 22, 108–26, 183
Herzig, Rebecca, 30
Holifield, Ryan, 111, 123
Humphrey, Caroline, 149
Hustak, Carla, 29

ice melt, 12, 18–9, 21, 27–8, 31, 33, 39, 40, 185
ice penetration, 33–4
iconicity, 186
imagination, 10, 11, 13, 15, 16, 17, 21, 23, 29, 87, 89, 90, 127, 128, 167; imaginative enactments, 114; resource imagination, 97
indeterminacy, 14
indexicality, 186
Indonesia, 81, 103, 128
Ingold, Tim, 22, 29, 69
instability, 38, 90
International Monetary Fund, 130
Italy, and environmental politics, 163–5

Kapferer, Bruce, 149
Keller, Eva Fox, 29
Kirsch, Stuart, 69
Klein, Naomi, 178
Kneas, David, vi, 13, 19, 22, 67–86, 185, 186
knowledge, 12, 19, 20, 46–8, 50, 51, 61, 108, 109, 110, 111, 112, 122–3, 147, 148, 164, 165, 177; anticipatory, 183–4; cultural, 94; dissemination, 17–8; expert, 29–31, 89, 95, 103, 108; geological, 74, 78; limited, 37; local, 30, 117; scientific, 33, 34, 39, 40, 41, 48, 56, 62, 116
Kosseleck, Reinhart, 164–5, 166, 178

Laidlaw, James, 148, 178
Lakoff, Andrew, 114
Latour, Bruno, 33, 94, 186, 187
Le Bon, Gustave, 133

Limbert, Mandana, vii, 17, 20, 22, 113, 120, 128, 147–62, 185
Locke, John, 132

Marx, Karl, 16, 143
material engagements, 11–2, 21
Mathews, Andrew S., vii, 9–26, 181
McClintock, Barbara, 29
Mexico, and silver, 21
mineral resources, 82, 83, 113; types, 73–4
mining companies, 13, 19, 67–86
Mitchell, Timothy, 95
model-building, 12, 21
modelled futures, 11
modellers, 11, 13, 18, 21, 22–3, 27, 30, 35–7, 41–2, 56, 62; and glaciologists, 35–6; and political environment, 13, 18, 22
models, 11, 12–3; cascades, 51–2; general circulation models, 12, 51, 52, 53–5, 56, 57, 60, 184; hydrological, 52; ice sheet models, 28, 35–7; limitations, 36–7
modernization, 16, 167
Myers, Natasha, 29, 30

nationalisms, 15, 30, 160; and futures, 15
nations, and Antarctica, 31, 43; and science, 30–1
natural gas, 11, 95, 155
NGOs, 130, 136, 142

O'Reilly, Jessica, vii, 22, 27–45, 183–4
ocean acidification, 9
oil, 9, 10, 11, 15, 16, 18; and Antarctica, 34; and nationalism, 160; and Oman, 17, 20, 147–62, 185; and São Tomé e Príncipe, 20, 127–46; depletion, 147–8, 155–6; discovery, 149–54; religio-political environment, 149–54
Oman, 17, 20, 147–62
Onneweer, Maarten, 17

performativity, 68, 69
Pinder, David, 167
privatization, 90
prognosis, 112, 168, 176–7, 178; as diagnosis, 68, 76–82; as prediction, 68, 71–6, 82; past, 163–80

prognostic politics, 9, 10–8, 70, 89, 164, 181, 182, 186, 187
pseudo-environmentalists, 79
public participation, 108, 109–10; and scientific risk assessment, 111–3; spokespersons, 115–8

Raffles, Hugh, 30
rainfall, 11, 18, 46–66
Rajak, Dinah, 69, 70, 77
religion, 9, 10, 150; conservative Evangelicals, 10
research stations, 30–1
resources, affect, 127–46, 184–5; environments, 97; imagination, 97; materiality, 68, 69, 128
Ricardo, David, 139, 144
Richardson, Tanya, 97
risk, 11, 12, 18, 19, 23, 92, 96; analysis, 14–5; and science, 22; in retrospect, 163–80; management, 10, 12, 15, 34; models, 13; non-quantifiable, 15; quantifiable, 15; reputational, 15
risk assessment, 14–5, 38, 108–26; and 'perpetuity', 120–1; boundaries, 108, 110, 122

Samimian-Darash, Limor, 182–3
São Tomé e Príncipe, 15, 21, 127; and oil, 20, 129–46
scarcity, 97, 98, 99
scenarios, 12, 13–4, 109; mining, 14, 113–5; planning, 13–4
Schechner, Richard, 70, 74
Scheiffelin, Edward, 68
scientists, and field environment, 27–45
Scott, James C., 16
semiotic ideologies, 94
sensory engagement, 19, 27, 28, 29–30, 37, 39, 41
signals, and uncertainty, 51–62
signs, 89, 92, 94, 102; material, 185–6
Simone, AbdelMaliq, 166–7
Skrydstrup, Martin, 34
Smith, Adam, 139, 144
Soares de Oliveira, Ricardo, 129
Spouge, John, 38
statistics, 10, 18, 56; credibility, 18

Strathern, Marilyn, 22
subjectivities, 29

Tarde, Gabriel, 133
temporalities, 10, 16, 20, 21, 82, 120; and
the city, 165–8
Toronto Stock Exchange, 69, 70, 71, 76;
rules, 69, 70, 73, 74
transnationalism, 92
transparency, 90, 99–100, 101, 130, 135
Tsing, Anna, 29, 69, 71

uncertainty, 12, 14, 20, 46–66, 121–2; and
anthropology, 47; and climate change
impacts, 56; bandwidth, 58–61; in the
signal, 51–62; scientific, 48; *see also*
risk
United Nations Development
Programme (UNDP), 138, 139, 140

US Environmental Protection Agency, 14,
19, 108–9, 110, 111, 112, 113–22, 123
Vatican, 140
Vaughan, David, 38

Walsh, Andrew, 131
water, 9, 10, 18; and divine power, 149, 159;
drought, 159–60
wealth redistribution, 140
Weber, Max, 16
Welker, Marina, 69–70, 81, 103
Weszkalnys, Gisa, vii, 11, 15, 20, 22, 97,
127–46, 178, 184
workshops, 50, 74, 79, 81, 149; theatre, 70,
74, 80, 81, 83
World Bank, 95, 96, 130, 134
Wynne, Brian, 11, 110, 112, 117, 122

Zeiderman, Austin, vii, 12, 18, 20, 22,
163–80, 182